Intelligent Systems Reference Library

Volume 230

Series Editors

Janusz Kacprzyk, Polish Academy of Sciences, Warsaw, Poland

Lakhmi C. Jain, KES International, Shoreham-by-Sea, UK

The aim of this series is to publish a Reference Library, including novel advances and developments in all aspects of Intelligent Systems in an easily accessible and well structured form. The series includes reference works, handbooks, compendia, textbooks, well-structured monographs, dictionaries, and encyclopedias. It contains well integrated knowledge and current information in the field of Intelligent Systems. The series covers the theory, applications, and design methods of Intelligent Systems. Virtually all disciplines such as engineering, computer science, avionics, business, e-commerce, environment, healthcare, physics and life science are included. The list of topics spans all the areas of modern intelligent systems such as: Ambient intelligence, Computational intelligence, Social intelligence, Computational neuroscience, Artificial life, Virtual society, Cognitive systems, DNA and immunity-based systems, e-Learning and teaching, Human-centred computing and Machine ethics, Intelligent control, Intelligent data analysis, Knowledge-based paradigms, Knowledge management, Intelligent agents, Intelligent decision making, Intelligent network security, Interactive entertainment, Learning paradigms, Recommender systems, Robotics and Mechatronics including human-machine teaming, Self-organizing and adaptive systems, Soft computing including Neural systems, Fuzzy systems, Evolutionary computing and the Fusion of these paradigms, Perception and Vision, Web intelligence and Multimedia.

Indexed by SCOPUS, DBLP, zbMATH, SCImago.

All books published in the series are submitted for consideration in Web of Science.

Yukio Ohsawa

Editor

Living Beyond Data

Toward Sustainable Value Creation

 Springer

Editor
Yukio Ohsawa
Department of Systems Innovation, School
of Engineering
The University of Tokyo
Tokyo, Japan

ISSN 1868-4394 ISSN 1868-4408 (electronic)
Intelligent Systems Reference Library
ISBN 978-3-031-11595-0 ISBN 978-3-031-11593-6 (eBook)
https://doi.org/10.1007/978-3-031-11593-6

This Springer imprint is published by the registered company Springer Nature Switzerland AG
The registered company address is: Gewerbestrasse 11, 6330 Cham, Switzerland

Foreword

With regard to the prevalence of technologies used with data, such as artificial intelligence and digital transformation, it is believed that digitalized information certifies improvement in various business and living sites. However, this book ambitiously aims to tell humans should and can acquire more useful knowledge than that acquired from available data, that is, to go beyond data. For "doing" innovations, i.e., for creating humans' activities to realize the latent value underlying data, the editor Yukio Ohsawa, who has obliged us by presenting keynote addresses in our KES International conferences, and the authors of this book highlight the reasons and methods for interpreting the hidden knowledge and using it to make new actions of people. Ideas created by thoughts based on the knowledge and experience of various participants in the market of data should be and can be connected via their interaction with other individuals and the real world by putting the ideas into real actions. Information regarding such actions is also converted into new data, and this conversion is accelerated. But let us ask… can this growth of data cover up all the ideas, knowledge, and experiences of all the people in the world? It would be great if the answer is yes—tools with AI will promise a happy life. However, if the answer is no, can we guarantee that all the actions and ideas behind actions can be shared and used to satisfy the requirements in the society? The more we ask, the deeper we have to discuss answering the questions, which urges us to ask new questions about the future of society with growing data. Thus, we encounter the necessity of methods to utilize information outside the data and the importance of human communication as a way to acquire useful knowledge. In addition, due to the spread of COVID-19, people became accustomed to digital communication platforms and realized that the use of emails and social networking services were just some ways to communicate online. Everyone is now witnessing the diversification of communication media, by which we can collect various data on the communication and interaction of humans. However, the design and creation of useful new data remain unexplored. These problems were addressed in the present study. The editor points out that there is a lot of information used without necessarily being in the form of data, and some information that is or is not digitalized has not yet been used. The research questions shared in all chapters here include "What kind of information can humans externalize and use to improve their quality of life?" and "What kind of information should be converted into data?" by answering which I believe living beyond data enriches data in machines, the life of humans, and happiness in society. Professor Dr. Ohsawa, the editor of this volume, is visionary, intelligent, and an inventor of new concepts. He is a dreamer of dreams. He created a new domain called chance discovery, meaning to discover events of significant impact on decision-making, in the year 2000. Since then, his original concepts and technologies have been published as books and monographs by world-class publishers such as Springer Verlag. I congratulate Prof. Dr. Ohsawa for evolving this book on my invitation in my capacity as a Co-Editor of the Springer Book Series entitled Intelligent Systems Reference Library. I am confident that this book will serve a large section of our researchers, practitioners, students, and professors.

Adelaide, Australia   Prof. Lakhmi C. Jain
May 2022   Ph.D. I ME I BE(Hons)I Fellow (Engineers Australia)
                                                   Founder KES International
                              http://www.kesinternational.org/organisation.php

Preface: The Reason Why We Seek Living "Beyond Data"

Living beyond data does not mean to live without data, but to acquire more useful knowledge than available from the data in hand. Here, more useful knowledge may mean knowledge that is more useful or a larger amount of knowledge that is as useful as available from data. The utility of information or knowledge may be defined in various ways. For example, in Chap. 1, the author regards the effect of information on the connectivity between the thoughts of stakeholders (i.e., the requirements of users, ideas of data scientists, and the knowledge providers of data and all others in the market of data) that plays an essential role in humans' decision-making and problem solutions.

This idea can be associated with the diffusion of innovation since Rogers' theory (1988, 2003) since it involves various stakeholders of the created values and the opened market. Here, not only the creators or developers of new products but also users play an important role to discover a new value of a product via using it and diffuse the value to the majority in the market. According to von Hippel (2006), leading consumers invent, not only use or diffuse, technologies. Thus, innovation can be regarded as the connected activities of people including individual thoughts, social communication, and the interaction of stakeholders in the market including consumers. This point distinguishes innovation from a child's talent of value sensing acquired in the growth of the mind presented by Donaldson (1988) or a part of sense-making that can be supported by information systems using data by Dervin (1988). That is, in the activities in innovation, the thoughts based on the knowledge and experience of various stakeholders of the potentially opening markets are connected via combining elements and doing the ideas in the real life. Efforts should be also dedicated to the reduction of the stickiness of information, as proposed by von Hippel, in order to foster these connections.

In today's society, the pace at which information about such doing is being converted into data is accelerating. In every field, data is being collected and accumulated. Let me give you an example to illustrate this point. In retail marketing, Position of Sale (POS) data used to be the king of data. It is no wonder that marketers think that it contains important information for their market strategy, because it is very rich data that contains everything: what was bought, its category, who bought it, the price, the time, the name of the store, and so on. For example, it has been said that unexpected patterns are learned, such as "people who buy diapers buy beer," and there has been a lot of discussion about whether this is meaningful knowledge or not. The background of this pattern was a situation in which the husband of a family who drives in on Sundays to buy a package of disposable diapers like pumps buys beer along the way. In reality, the story is a figment of the imagination, but in any case, many people have come to realize that patterns derived from data through machine learning techniques only make sense when interpreted by humans. In this case, instead of selling diapers and beer, the best way to attract customers might have been to play music in the pump store that would soothe them after a long day of work.

I as the editor and the author am not a marketer, but as a data scientist, I was fascinated by data from markets in 2000 when I first encountered it. The example described in Chap. 1 is based on sample pickup data of textile products, not POS data of a supermarket, and visualized with KeyGraph, which I developed in 1996. KeyGraph is a tool for visualizing (1) items that appear frequently (black nodes), (2) clusters (islands) consisting of lines (black links) that connect items bought by the same person at the same time with high frequency, and (3) rare items (bridges = red nodes) that co-occur with items on multiple islands. Initially, I expected that red nodes appearing in KeyGraph would be promising products in the future. The reason for this expectation was the optimistic belief that the islands formed as in (1) and (2) would be existing hot-selling products, and the bridges obtained between them as in (3) would be hit products that would connect the hot-selling products and create new hot-selling products. For example, there was a cell phone called i-mod in Japan at that time, which was an epoch-making hit product then because it combined the functions of a telephone with those of a PC, such as e-mail and Internet access. When I visualized the human network from the conversation logs of SNS (in this case Yahoo), I found that the red nodes tended to show innovative people.

However, this expectation was not right. Some of the products that appeared at the bridge would subsequently increase in sales, while others would decrease, and which of these would be the case was not statistically significant. Even if the product on the bridge is a new product that was born from multiple hot sellers, has to climb a considerable hill before it is accepted by the market—new ideas from innovative people are rarely accepted because of their novelty. And, in order to improve sales in the market, it was necessary to go through this "acceptance" process. However, the fact that we could not create a method to extract products that would increase sales here does not mean that there are no potential business expansion opportunities. We can expect opportunities, as long as one believes there is something to be gained from such data as I had at the time. This expectation can be outlined in a non-descript linear equation like the one below. What we get from data is not only patterns that are learned from inside the data. It is also the case that knowledge of intrinsic value cannot be acquired without piecing together the usually disparate fragments of human empirical knowledge that lie outside the data.

$$\Sigma : \text{Knowledge needed to live} = \xi(\text{patterns in data})$$
$$+ \phi(\text{empirical knowledge outside data})$$
$$+ \zeta(\text{the component that connects the elements in } \xi + \phi) \quad (1)$$

So, how do we create and combine all of this? As an example of taking advantage of such diversity and putting it into practice as textile marketing, as I mentioned in my papers and in my book Innovators Marketplace from Springer Verlag (2012), and in this book I added another interpretation to this result in Chap. 1, patterns in the data are obtained by learning and visualizing the structure of the islands in the cluster and the bridges between them from the data on product selection by the customers corresponding to the POS in the business. Although the relationships as the patterns are statistically significant because of the high frequency of the island interiors, they are not significant for the bridges because of the low frequency. From the perspective of traditional statistics, the latter kind of relationship should have been removed on the grounds that it was unreliable. However, our idea of chance discovery (see Chap. 1) was that if a certain pattern or knowledge is "unreliable," we should think about "under what conditions it can be trusted," and use the pattern as knowledge after satisfying the conditions. In this case, the bridge is a way of asking what conditions (i.e., "why") some customers wanted a product that not many customers want very often. The bridge in the textile example was corduroy textile, and the motive of the customer who wanted it was that female office workers need clothes that are easy to wear when they go out for dinner at night. Furthermore, office workers need warm clothes to go out at night, and most importantly, they need clothes that they can relax in instead of wearing a suit at work, and if the situation calls for it, they need clothes that are fashionable. This may seem like a wild guess, but it is the product of years of hard work by marketers observing youth fashion at work. Those years are often even longer than the available market data that has been accumulated. Thanks to this, we can hypothesize the "conditions" that lie behind the data. Yet, if it were not for the communication that went into sharing the data visualized here, there would have been nothing more than disparate and fragmented dissemination of empirical knowledge—the friendships of women who go out to dinner at night and the warmth of their clothing. The result here was to link these through communication, where one person's opinion was relayed by another.

Although we invented the processes of chance discovery, they were not always easy to realize. In all the process models, the key point is not only to learn or visualize some pattern from the data, but also to execute the metacognitive process as an important part to materialize and structure the findings. Meta-cognition is "cognition about cognition" (meta-X means X about X, just as metadata is data about data). This is achieved in the process of striving to verbalize or formalize the tacit knowledge of which the bodies have as unspoken experience. Furthermore, by conducting such a process not only alone but also with colleagues in the workplace, we can become aware of the relevance of connecting various experiences.

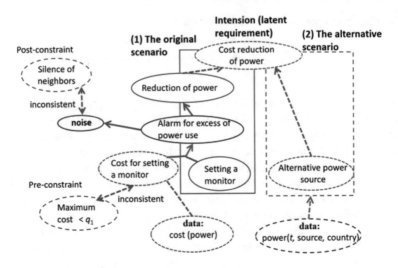

Fig. 1 A causal tree for linking requirement and data for power-generator state management

For example, suppose two people want an orange, but there is only one orange. One person asks, "Why do you want it?" The other person replies, "To eat it" and asks the same question, the other says, "I want to make marmalade." Then they ask each other, "How?." One wanted the pulp and the other wanted the rind. As they repeat the question and answer, they gradually shift their attention from the orange itself to their own motivation for action and the process of realizing it. This kind of meta-cognition through communication is also useful in more practical situations. Figure 1 shows a similar causal tree for the state management of an electric power generator. In the solid line on the left half of the figure, a mechanism to introduce sensor networks and other devices to manage stability and cost in power generation is considered, but gradually the severity of cost and other constraints is realized. Eventually, they realized the limits of their previous thinking and realized that they needed a new way of thinking, so they decided to rethink the power generation system. In this way, shifting attention through meta-cognition works effectively to counter the question, "Why are we choosing this method in the first place?

The necessity of information outside of data and the importance of expression and communication as a way to shift attention and grasp useful information are described above. Then, how do we utilize and reuse the objects of our attention that we have deepened in this way? If we are to reuse it, we need to accumulate the changes in our attention as useful information, but is this something that can be obtained in the form of data? Ten years ago, the most trendy data in the field of artificial intelligence was the history of statements and photos on social networking sites. These were also useful as tools for analyzing human minds and detecting or predicting changes in them, and were therefore used extensively in marketing and policy initiatives. On the other hand, social networking sites have also been used to predict stock prices and explain the background of changes. It can be said that this in itself could be considered as something distinct from quantitative analysis, which relied on the old chart data. Later, however, people in all fields came to their senses and realized that the use of social networking services was just one of the ways in which people were trying to get out of the world of classical data used in their own fields and use new data. With the diversification of data types, people are noticing their behavior is not only limited to the movement of things outside the body, such as the time series of products purchased, and positioning some of each person's thoughts as stored on social networking sites. People's movements are collected via smartphones, and in addition, data related to their physical condition is combined with advances in artificial intelligence technology. However, research on how to search for and obtain new data, or how to conceive, design, and create new data, is still unexplored.

Currently, there is a lot of information used without necessarily being in the form of data. For example, many middle school and high school swimmers remember how they swim by feel, and try to replicate that feeling during warm-ups before practice or races. If the feeling of being able to swim at maximum speed returns during the warm-up, they will feel at ease going into the race, but if it does not, they will stand on the starting blocks with anxiety. So, even before the warm-up—for example, the night before the race, at bedtime, with their eyes closed—a swimmer tries to revive the latent race feeling in their body by holding the posture of their body straight in the water, facing the direction they are going in, on their bed, or visualizing oneself swimming. He tries to revive the potential feeling of the race in his body. However, if he has a video of oneself swimming, even if he just watches it and does this image training, one will not only recreate the sensation of the race but also improve swimming technique through meta-cognition. Furthermore, by using technology to visualize the difference between himself and the best swimmers, the process of improving swimming technique can be made even more efficient. Using current artificial intelligence technology, it would be possible to extract and visualize the essential techniques that are lacking in his swimming. Thus, more and more of the latent know-how—tacit knowledge—used in a non-data form will be converted into data over time, bringing new insights. Obviously, however, the use of data is not to completely automate swimming and create a swimming robot. Since the goal is to improve human swimming, the information processing needs to be relayed to the human brain, trunk, and limbs. Hence two problems.

(1) What kind of information can human(s) externalize and use to improve their swimming style? (2) What kind of information should be converted into data as a prerequisite for the computer to output and show the information as in (1) above?

When written in this way, we can see that we have no choice but to discard some part of available information both when presenting information to people (1) and when inputting information as data into a computer (2). Relatively few people would include information about the building of a swimming pool when inputting data related to swimming techniques, and few swimmers would want to look at the ceiling of a swimming pool when the results of computer analysis and visualization are presented. In this way, the underwater movements of swimmers will be analyzed, and the results will be used by many swimmers to do close to the top swimmers. Then they realize that it is very difficult to mimic only the movements following the computer's instructions—if you try to swim in a pool by holding your body straight and moving your hands along the axis of your body. This is because the more rigidly you try to fix your body posture, the more your body sinks, which makes it unstable, and the more you follow the trajectory of your hands with your eyes, which makes your body shake even more. In the case of crawl stroke, the eyes should look slightly ahead of the bottom of the water, and in the case of backstroke, it is better to look at the ceiling or the rope attached in the direction of movement. When one realizes this, the structural features of the pool that were discarded from the data earlier, such as the lines on the bottom of the water and the construction of the ceiling, are taken back as data. In other words, by repeating the cycle of using data to gain knowledge, using the knowledge to take action, acquiring new points of attention through the action, and collecting data, people update their behavior, knowledge, and attention revision, and collect data. By repeating the cycle of acquiring new points of attention and collecting data, we update our behavior, knowledge, and attention (attention revision), and update our data (collection). In other words, one's knowledge grows beyond the data at hand, and one's behavior evolves to include actions that do not rely on prepared data. It is this process that this book deals with.

In Chap. 1 by Ohsawa et al., the authors introduce the problem of the gaps among requirements and the use of datasets and AI tools. Feature concept is shown as a tool for linking them, to enable to use and combine datasets needed for innovations. FC works effectively as a way to connect social demands, ways to meet them, and data, and is used to advance human thinking and communication by overlaying contextual information that does not appear in data. Furthermore, the creation of the data to involve in the process of data mining is the challenge of FCs. In Chap. 2 by Auernhammer and Leifer, this thinking is further generalized to show the importance of "why" questions in designing things and systems, i.e., reasons for a design, and "how" questions, i.e., the method by which they are used in actual system design. The fact that the "why" question is at the heart of the process of forming and selecting the perspectives necessary for innovation will be essential knowledge for those who want to break away from dependence on existing data and achieve innovations based on truly necessary information. In Chap. 3, Magnani situates the significance of why questions for living beyond data by introducing the philosophical tool of abductive cognition. Abductive cognition, having "why" questions as the key factor, refers to all kinds of reasoning to more or less deep hypotheses starting from data. He proposes locked and unlocked strategies that shed light on the cognitive aspects of both humans and machines, related to the outputs of abductive cognition ranging from weak (closer to data) to strong level (beyond data) of knowledge creativity. In Chap. 4, Kondo, with the editor Ohsawa, shows cases of workshops executed in the living labs in local regions of Japan. She points out the effect of a living lab is, rather than direct acceleration of innovations, to revise conceived links between the contexts of participants' lives. This revision of the links means to disconnect existing links and connect missed links with the openness of the minds of participants. Readers may find this idea has a similarity to Magnani's cognition "out" of data by unlocking the human mind where minds are released from the prejudice to adhere to data in their hands and get linked to new information sources which may be other individuals.

Above is the first part. The second part is about the general process to explore, collect, and use data for living beyond data. The use of data includes the explanation of why/how to use data, to provide readers with the motivation and methodology to go beyond the existing data-based innovation/thought support methods. Chapter 5 by Tsang and Benoit points out the problem of bringing up post-hoc explanations to explain black box AI from a machine learning perspective. In the case of explaining complex and difficult-to-interpret models such as deep learning, in the words above, we are explaining the behavior of the system, which is the how, and not the movement of the target world, which is the why. In Chap. 6, GAM models are evaluated for sepsis prediction in terms of interpretability and predictive performance. The shown GAM models offer a balanced trade-off between explainability and predictive performance toward users. We put these two chapters at the beginning of Part II as a way to raise the issue, because explanation and prediction of complex systems are essential for bringing about innovation in the target world. In Chap. 7, Yamamoto and Kondo emphasizes the importance of the process of data mining not only the sheer knowledge about mathematics, data science, and AI, which students are urged to learn in the recent social trend including elementary level education. By presenting a process including data acquisition and data management, both educators and learners can design what should be learned by clarifying the goal of using data and of the education that is open to the living of humans in the real world, i.e., beyond data. His comparison of data-science processes with software development models clarifies the possibility of created processes. In Chap. 8, Abe reviews the framework of abductive reasoning, which leads to an explanation of machine learning corresponding to answering the question "why is such a result obtained?" Because the author is an expert of the framework of analogy-based abduction, the ideas shown here inspire ways for importing knowledge from external domains that have been desired so far in the real application of AI tools.

The third part presents insights from exemplifying ways to go from using data to living beyond data. Chapter 9, by Bandini, Gasparini, and da Silva, is about the behavior of people and their perception of risk from sensors in order to achieve a walkable city. The analysis method used here is inherently interpretable, so the results can be interpreted immediately, so that policies to realize people's happiness in urban design shall be proposed. It can be seen that the condition of being inherently interpretable is not only for explainability, but also for the reasonable development, selection, and use of necessary sensors. Chapter 10 by Sakamoto and Nozaki further explores the essence of living beyond data. In this chapter, she developed an AI system that generates new mimetic words based on an experiment in which she decomposed the Sound Symbolic Word, which expresses the feeling of the texture of various object surfaces in Japanese, into syllable combinations and related the expressions of mimetic words to subjective sensations such as the feeling of comfort felt by people. This in itself is a method that goes beyond AI for handling existing linguistic data, but the author also proposes the development of new values and materials through the interaction of people with new mimetic words, which is the definition of innovation proposed by Drucker. Chapter 11 by Shimokawara focuses on natural language, but what it achieves is a system of conversations between robots and people.

When the robot re-asks the user's answer to the robot's mundane question about the seasons, the user gives a variety of answers that reflect his or her own senses. This is not a particularly complex new algorithm for machine learning or data analysis, but a way to break out of the templated data description format and put in essential human knowledge. Chapter 12 by Fujii et al. presents the method for and results of extending data by introducing the idea of data-assimilation for a strategic designs of a traffic system as an example, with two approaches. The first is the forward simulation for generating higher resolution data to go beyond data collectable from the real world, and the second approach goes in the reverse manner using the output from the simulation and partially observed data in the real world to estimate the hidden factors behind the phenomena observed in the target system. The reliability of these approaches depends on the reproducibility of the simulator, so the author Fujii to apply inverse analysis where link traffic volume is chosen as real world data and used iteratively to fit the model to social demand. This point is connectable to the consideration of alternative systems in the stakeholder communication introduced in Part I. Finally, in Chap. 13, Hayashi and Ohsawa address the problem that questioning what data are required for any purpose is insufficient related data sharing. Data are principally provided by organizations who publish the data unilaterally—not just one-way free flow rather than data exchange toward innovations with eliciting and sharing users' requirements. To address this issue, they introduce the concept of "data origination," the act of designing/acquiring/utilizing data reflecting subjective knowledge and diversity of perspectives of humans. The case study of this idea is shown about data externalization for suppressing the COVID-19 pandemic.

In summary, living beyond data enriches data and humans' thoughts than discounting the value of data. Let us not go longer here. Readers are kindly urged to open Chap. 1.

Tokyo, Japan   Yukio Ohsawa
January 2022

Contents

Part I
Thoughts and Communication for Living Beyond Data

Chapter 1
Living Beyond Data with Feature Concepts

Yukio Ohsawa, Sae Kondo, and Teruaki Hayashi

Abstract The inconsistencies between the requirement for, the collection of, and the use of data, and the inconsistencies among analysis models for dealing with different datasets have been solved via communication and thoughts using data jackets to solve problems in real life and solutions to problems using data. In these cases, essential information for the data-federative process has been externalized and used/reused. Feature concept, a model of the concept to be acquired from data that cannot be represented by a simple feature, such as a single variable but can be by a conceptual illustration, turned out to be a core essence here. Decision trees, clusters, and even deep neural networks can be regarded as examples of feature concepts. Useful feature concepts for satisfying the requirements of a data user have been elicited so far among data providers, data users, and other stakeholders in the market of data. In this chapter, the data-jacket-based design of creative communication is reviewed with some cases of application—marketing, detection of earthquake precursors, suppression of COVID-19, cases, and highlight the elicited feature concepts.

Keywords Innovation · Data utility · Feature concept · Data jackets

1.1 Introduction: Communications for Discovering Data Utilities

So far, data have collected the attention of people living in environments where data have not yet been collected. In contrast, many people working in businesses collect data, although the reasons for or the benefit of using data have not been clarified. These gaps between the requirement for, collection of, and the use of data are to be solved by living beyond data because the inconsistencies come from the lack of

Y. Ohsawa (✉) · T. Hayashi
Department of Systems Innovation, School of Engineering, The University of Tokyo, Tokyo, Japan
e-mail: ohsawa@sys.t.u-tokyo.ac.jp

S. Kondo
Department of Architecture, Graduate School of Engineering, Mie University, Tsu, Mie, Japan
e-mail: skondo@arch.mie-u.ac.jp

© Springer Nature Switzerland AG 2023
Y. Ohsawa (ed.), *Living Beyond Data*, Intelligent Systems Reference Library 230,
https://doi.org/10.1007/978-3-031-11593-6_1

knowledge and meta-knowledge, that is, knowledge about knowledge that may come from outside of available data. By creating such knowledge, the utility of a dataset can be explained by its relation to information from the data in hand, including unknown datasets. However, so far, we found that information obtained from a combination of connectable (via sharing attributes) datasets from different domains tends to be difficult to interpret because the analysis model for a dataset may be useless for another dataset. There is a gap in scientific explanations in different domains that cause difficulties in data-federative innovations. Thus, we need a method for urging users of data to be aware of the gaps in the vertical direction, that is, requirements versus data, and in the horizontal direction, that is, between the real worlds where data are collected. The gaps should be solved if possible for creating connections in the two directions, whereas the connection is not a good choice, if impossible.

To discover the utility of data, collecting stakeholders - data providers, data users, data workers, and scientists—and facilitating communication is essential for solving the gaps mentioned here. The essence of this communication includes at least three steps that may occur in various orders. The first step is to elicit the requirements of data users who may not touch the data directly but may receive the services or products created and provided using the data (the top node in Fig. 1.1a). In the second step, the workers who touch and use the data directly will invent a method called the solution to acquire knowledge to satisfy the requirement via a process to use statistical or AI-based tools, so that satisfactory products or services can be obtained (the right-hand node in Fig. 1.1a). Then, in the third step, the worker should collect the dataset useful for the solution invented in the second step by calling for suitable data providers (the left-hand node in Fig. 1.1a).

Because this is a human-to-human communication that affects the benefits the participants, explanations of reasons for proposing the solutions are desired as a responsibility in business. For example, the customer living in Osaka who speaks about his requirement to sustain his physical health should ask why a service to propose an everyday meal has been provided. If the answer from the data worker refers to data on blood pressure and diseases of citizens of Tokyo, used in providing the service, the customer should ask why the data from Tokyo instead of Osaka is regarded as useful. To answer this question, the worker must prepare by asking the data provider "why can this data from Tokyo be useful for managing the health of people in Osaka?" to explain that the subjects' conditions of living in Tokyo were similar to Osaka, referring to the way the data were collected. Such communication has been realized since 2013 in the Innovators Marketplace on Data Jackets (IMDJ) [26], borrowing the basic idea of innovation support from Innovators' Marketplace [24], which realizes a part of the process for innovation, defined as creating new dimensions of performance (as in Drucker's definition of innovation [6]), because new requirements came to be satisfied using the datasets. As shown in Fig. 1.2, a data jacket (DJ) is metadata for each dataset, reflecting peoples' subjective or potential interests. By visualizing the relevance among DJs, participants in the market of data think and talk about why and how they should combine the corresponding datasets.

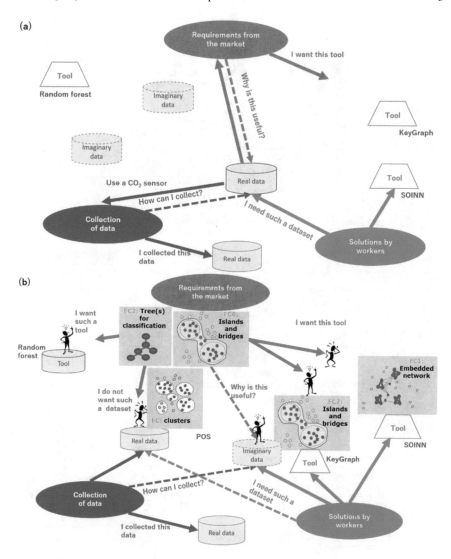

Fig. 1.1 The communication among the users of services created by use of data, who tell requirements, collectors of data, and workers who directly touch and user the data. As in **a**, a requirement tends to be hard to meet satisfactory data or tools (vertical gap)and the data are also hard to meet other data (horizontal gap). As in **b**, feature concepts are useful, if elicited, for reinforcing the connections. We shall return to feature concepts in **b** after referring to Fig. 1.2 to see how communication has been realized so far

(a)

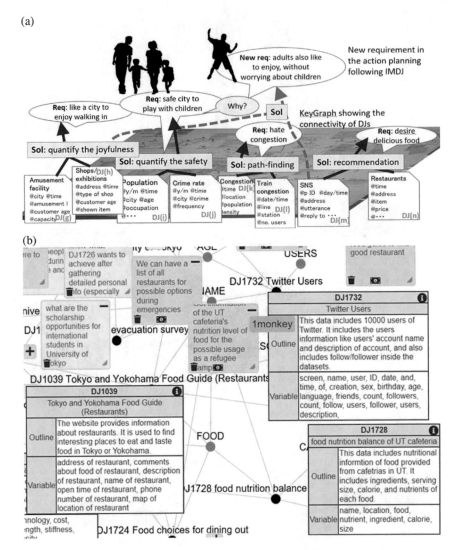

(b)

Fig. 1.2 A snapshot of IMDJ, **a** face-to-face and **b** on-line. Referring to the visualized map about the connectivity between datasets using DJs, participants communicate requirements corresponding to the externalization of problems and solutions to the problems as in **a** where Req and Sol respectively mean a requirement and a solution. The ideas about solutions, where datasets are used in combination, can be proposed easily because not the contents of data but small pieces of information summarizing the datasets and their utility are provided in the form of a map connecting DJs visually as in **b**. Solutions (squares, e.g., "We can have…") should be proposed by combining DJs (large cards e.g. DI1039) responding to requirements (e.g., "what are …..") as in **b** for data-federative innovations

Even if the owners of data may hesitate to open their data to the public due to the consideration of opportunity loss or privacy, they can present only the DJs to the public society because DJs do not include the content of the data. In this sense, DJs are tools for fostering creative communication and protecting privacy, and IMDJ is a place for realizing the market of data, i.e., a platform for innovations using data. Here, participants communicate to elicit ideas for combining/using/reusing data or future collaborators. Furthermore, explicitly or implicitly required data can be searched by use of tools developed on DJs [11], which enabled so far, for example, analogical inventions of data analysis methods based in a previous solution to solve a new target problem. In the cases presented later, data-federative (i.e., combining data) innovations came out from the IMDJ.

1.2   Feature Concepts for Connecting Stakeholders

We reviewed the process of IMDJ by modeling it on the logical framework of abduction to represent thoughts and communication [27, 29]. Thus, the necessity to elicit information about contexts, which means the conditions for using data and receiving the services or products created based on data, has been clarified in the form of a theory that supports a solution invented to satisfy a requirement. Simply put, a context refers to how a solution can be realized and/or why a requirement has been proposed. For example, if the requirement is to reduce traffic congestion, the reason (answer to "why") may be relevant to multiple contexts, for example, desiring to avoid infectious pandemics or to live in a walkable city free from traffic accidents and crimes. If the context is the former, the datasets to be used in the solution may be about human flow, population, effective reproduction rate, etc., whereas the data should be crime history and human flow if the context is the latter. Contextual information is essential in embodying a data-based solution to realize a requirement, and asking deep reasoning questions (why) and generative design questions (how) contribute significantly to data-federative innovations as well as designing [7]. The limit of IMDJ comes from the gap between the expressed requirements reflecting humans' subjective desires or thoughts and the data to be used which do not reflect the subjectivity of stakeholders in the data market. Although DJs accept subjective descriptions by data providers or individuals with knowledge about data, they are still insufficient because of the bias of individuals' interest compared with the variety of interests in the entire society. Therefore, we urge as many freshmen as possible of the IMDJ system to show subjective opinions to other participants in the IMDJ, so that we can reflect the entered opinions as a part of DJs in the future. However, we still suffer from a more important missed link between data and the requirement, that is, the feature concepts introduced below. A feature concept is a conceptual representation of the information or knowledge to be acquired by using data, which is linked to the method of how and why a dataset(s) should be used and to satisfy a requirement. In the examples shown in the next section, we discover that creativity in IMDJ has been ignited by and for externalizing, using, and sharing feature concepts

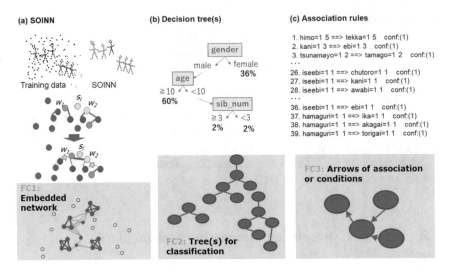

Fig. 1.3 Examples of feature concepts (FC1, FC2, FC3) for self-organizing incremental neural network (SOINN), decision tree(s), and association rules

in various forms. The form of a feature concept could have been a new variable, a new predicate, a new function, or even a new logical clause if the participants could express all concepts in the logical framework of abductive reasoning, but it is more likely to be illustrated as an image as "FC" in Fig. 1.3a. The no-supervised machine learning method SOINN [10] is an algorithm, where, briefly put, existing samples close to each new sample are connected by a new or a strengthened edge whereas other edges are weakened. The result is the restoration of a hidden structure from scattered noise signals. The embedded network structure, assumed for this restoration, reflects the requirement to extract the latent structure from a noise-rich image and can be interpreted as a feature concept of SOINN.

Various basic methods for machine learning can be classified based on feature concepts rather than algorithms. For example, as shown in Fig. 1.3b, decision tree learning is used to extract trees [30, 31] for classifying all the entities in the data to explain a given class, and the random forest [13, 14] means extracting a forest that is a set of trees. In these methods, trees and forests are used as feature concepts, suiting the requirements for classifying samples and obtaining multiple trees for a more reliable classification than a single tree. If a forest is not intuitive to a user, such a feature concept as a conference can be better used where the idea of each participant in a conference corresponds to a tree. On the other hand, association rules to be obtained by the traditional Apriori method [1] can be represented by a feature concept, a set of nodes connected via arrows, where the nodes at the tail indicate the conditional items or events, and the nodes at the head indicate the conclusional ones that tend to occur if the conditional ones do. The algorithms presented in the above references in each bracket share the feature concepts, although the algorithms may vary owing to the various efforts of the researchers who developed the sophisticated methods. In these efforts, other feature concepts may have been exploited as parts of the feature concept at a higher level. For example, in the case of decision tree, the variables to be used at a higher level such as the gender ("sex" in the figure) of an individual to explain the survival rate in the Titanic accident is selected on the variable's quantity of information i.e., maximum reduction of the entropy since the earliest version ID3 to more recent random forest. This use of entropy can be interpreted as keeping the branch of the tree that separates positive samples (victimized individuals) from negative samples and cutting other branches. This branch may be regarded as a feature concept as a part of a tree, fitting the requirement to select a useful variable—a variable corresponding to the branch. Feature concepts, if provided explicitly, play an important role in bridging social requirements and features in datasets because they originate from the requirements of data users and can be projected onto the methods, tools, and algorithms to be used to take advantage of the . As shown in Fig. 1.3, a feature concept may be given by someone other than the creator of an algorithm or the collector of a dataset, if the purpose is to foster communication about the tool in the market of data. If a feature concept is given by a data user him/herself, the communication for reforming it to a computable procedure (algorithm or heuristics) turns out to be a process for requirement acquisition for the analyst and for metacognition for the user who is the client of the analyst. Thus, contexts and feature concepts connect elements of the data market—datasets, knowledge, and skills for using/reusing data—toward data-federative innovation, as shown in Fig. 1.4. As well as a frame of desired knowledge or of thoughts and communication about the knowledge, the

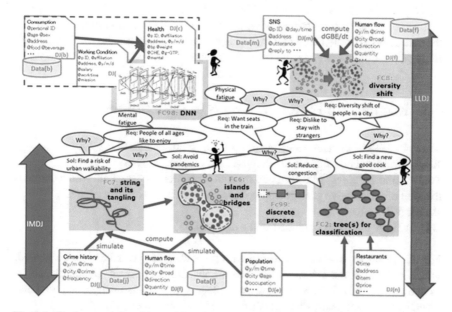

Fig. 1.4 The images and positions of feature concepts (FC2 through FC99 here) in the communication to connect requirements ("R") to solutions ("S") and DJs ("DJ()")

feature concept can be used as the performance dimension of data analysis because it is the feature desired to be extracted from data.

Here, let us provide a brief summary of the abductive reasoning framework for IMDJ as presented in [29]. DJ_i, the i-th among all presented DJs, is defined as a set V_i, F_i, P_i, C_i, W_i, where V_i, F_i, P_i, C_i, and, W_i represent the variables, functions defined over some variables, predicates defined over some variables, clauses composed of these elements, and words summarizing the corresponding dataset D_i including subjective expectations of the utility of the dataset. In IMDJ, the communication goes on by adding new predicates and functions that had not been covered by F_i and P_i, which means new dimensions of performance of a created services or products, adding words to W_i corresponding to contextual information. In addition, new variables to compose new predicates or new functions can be added as a new DJ. As a result, combinations of DJs and added elements are proposed as a solution for realizing goal G, corresponding to the satisfaction of a requirement, in a way of abductive reasoning. A context can also be represented by the intentions of or constraints on stakeholders, which can be represented by an element of either V, P, F, or C, as well as W written above. Here, a feature concept may be represented as an element of P (e.g., line(h_1, h_2) in one of the following examples), F (e.g., change(t), trend(t) and diversity(t)), or a clause (e.g., $trend_1(t{:}t{+}Dt) :\!- trend_2(t{:}t - Dt)$, change($t$) representing a change explanation) in C. Thus, the range of concepts to be represented by FC corresponds to various dimensions of the logical framework of abduction. In this sense, we should encourage participants in the IMDJ to externalize feature concepts, accept various ways of communication to make each participant's awareness reach out to others' contexts and to externalize and exchange feature concepts by combining Living Labs and DJs. As shown in Fig. 1.4, LLDJ is the combination of the IMDJ and living lab expected to increase the width and depth of the elicitation of contexts and feature concepts.

1.3   Examples of Feature Concepts in Data Utilities

Example 1 **Skill development in sports** (Req: requirement, Sol: solution).
Req1: Evaluate and improve the defense skill of a soccer team
Sol1: Visualize "*lines*" of teammates on which to quickly pass a ball, which explains the skill of a defensive team to manage the changes in the offensive team.
DJ1-1: Wide-view video
DJ1-2: Body direction (included in the data of DJ1)

For this example, an IMDJ session inviting coaches of sports has been executed, which resulted in the creation of a dynamic visualizer of "lines" in soccer games, as shown in Fig. 1.5. The lines between players are automatically computed and visualized for each team, distinguished by red and blue lines [32].

In this case, the solution proposed in the IMDJ was originally "*evaluate the defense performance of a soccer team based on the positions and body angles of players,*"

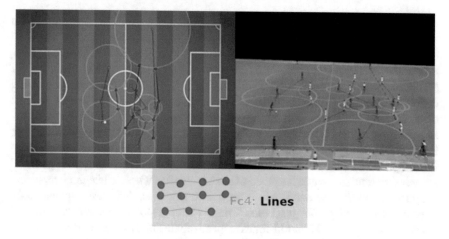

Fig. 1.5 The obtained product from an IMDJ session inviting coaches of sports, that is a dynamic visualizer of "lines" in a soccer game. The lines between players are automatically computed and visualized for each team, distinguished by red and blue lines [32]

which was then revised to a simple feature called "*lines*" of players. This revised feature can be expressed by four Horn clauses (6) through (9), where h_i for i of 1 through 11 represents the 11 players in a soccer team. The defense performance of a team is high if the players form three lines, where each line has three or four players positioned side by side in parallel to the goal line. The video data requested as DJ1 were eventually used to detect the lines formed by the players. The feature concept "line" is embodied here in the form of predicate "in_line" in clauses (1) through (3), which came to play an essential role of the connection between the goal (the predicate "defendable" in clause (4)) and the data (i of h_i for 1 through 11, taken from the video data). The lines were obtained by detecting each player in the video data by distinguishing the uniform clothes of the two teams, computing the body angles and the angles between the goal line and the line connecting each pair of players. Thus, "line" came to be a computable concept externalized via the conversation about available data and the latent requirement. Thus, the feature concept is a *line*. Figure 1.5 shows the created software reflecting the solution above, a visualizer of lines in the offending and defending teams from real video data of a soccer game. Using this tool, the soccer coach who presented the requirement above came to lead all his teams, none of which had been previously ranked within the top 32 in his prefecture before the IMDJ session, to be ranked within the top eight.

$$good\_defense(h_1, h_2, h_3) : -in\_line(h_1, h_2, h_3) \tag{1.1}$$

$$good\_defense(h_4, h_5, h_6, h_7) : -in\_line(h_4, h_5, h_6, h_7) \tag{1.2}$$

$$good\_defense(h_8, h_9, h_{10}, h_{11}) : -in\_line(h_8, h_9, h_{10}, h_{11}) \tag{1.3}$$

$$defendable(team) : -\forall H[good\_defense(H), in(H, team)] \tag{1.4}$$

The next example is not from an IMDJ session but from the development of KeyGraph in 1998 [25] by the author. KeyGraph was first devel-oped as a method for indexing, that is, extracting keywords from a document (D2-1 below). However, it is currently used as a tool for information visualization systems, typically for aiding the process of creative decision-making [15, 16, 19] from other sequential data (D2-2 below). At the time of its first development, the situation was the interest of natural language researchers at the time of extracting a rare keyword, that is, a low-frequency word that carries the essence of the contextual flow in the target document. For example, in Fig. 1.6, the visualization of word-word correlations is shown using KeyGraph. The story of The Arrest of Lupin changes the context from a peaceful journey on a tourism boat to the mood of anxiety due to a wireless telegraph reporting the information that Arsene Lupin is on board. However, "wireless" appears only three times and "telegraph" just twice, whereas "Lupin" the rubber and "Ganimard" the detective appeared 37 and 11 times respectively in spite of the small information they carry (Lupin and Ganimard appears quite usually in the novel series of Arsense Lupin by Leblanc). The problem was how to extract "wireless, telegraph," or the pairwise couple as the keyword of this text. Because "wireless" and "telegraph" are incomparably more frequent than "Lupin" or "Ganimard" in ordinary documents other than this novel, the tfidf criteria evaluates the importance of "wireless telegraph" even less than the sheer frequency.

The success of KeyGraph in Fig. 1.6 was because this algorithm was created by responding to the requirement and the idea (solution below) for the datasets as follows. Note that DJs had not yet been invented at the time KeyGraph was created, so let us put just D instead of DJ.

Example 2 **KeyGraph on the architecture model** [25]
Req2: Low-frequency words or items representing an essence in contextual flow in a text or sequential data
Sol2: Visualize "bases" and "roofs" in the metaphor regarding the target data (text) as an architecture of a building, and take the roof as the keywords or key items.
D2-1: text (document)
D2-2: Sequence (items in a supermarket, earthquakes: applied later than DJ2-1)

As a result, KeyGraph has been proposed based on a simple model called the architecture model (building construction metaphor), where a document is compared to a constructed building. This building has bases (statements for preparing basic concepts), walls, doors, and windows (functions to protect people inside). However, the roofs (new ideas in the document), without which the inhabitants of the building cannot be protected against rain or sunshine, are regarded as the most important and are supported by columns. KeyGraph has been regarded as a tool to find roofs, that is, terms that hold the rest of the document together via columns. To extract the roofs, KeyGraph is composed of three major phases.

(1) Extracting bases, where the basic and preparatory concepts are obtained as clusters constructed on the cooccurrence among the terms in the document.

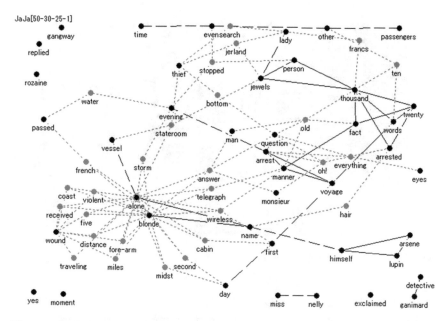

Fig. 1.6 Two feature concepts of KeyGraph

(2) Extracting columns, where columns, i.e., the relationships between terms in the document and the basic concepts extracted in (1) are obtained.
(3) Extracting roofs, where nodes (representing terms) at the cross of strong columns are regarded as roofs.

Making a co-occurrence graph in (1) was, already in 1998 when the author developed KeyGraph, a common approach for computing the relevance between terms. In contrast, the novelty of KeyGraph was to regard the co-occurrence graph as the basis of the document, by which the ideas in the document are supported. Thus, the result of KeyGraph consists of the following objects, unifying "word" and "item" used above into "item" in the descriptions below.

- Black nodes indicating the items in the bases that frequently occur in a dataset.
- White (or red in a revised version) nodes indicating that the *roofs* did not occur as frequently as the black nodes but co-occurred with clusters of black nodes.
- Double-circled nodes indicating items that co-occurred, especially frequently with the items in multiple clusters of black nodes. These are the most important roofs.
- Links indicating that connected item pairs co-occurred frequently. A cluster of nodes connected by solid lines form a *basis* (a cluster of co-occurring black nodes), whereas the dotted lines connect bases (connecting an item X and a cluster of items co-occurring with X) via *roofs*.

The *architecture* can be regarded as a feature concept. The difference between a feature concept and a model for analysis is that a feature concept comes from human ideas originating from thoughts about daily requirements, which may have to

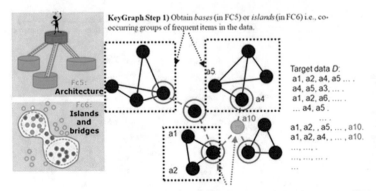

KeyGraph Step 1) Obtain *bases* (in FC5) or *islands* (in FC6) i.e., co-occurring groups of frequent items in the data.

Target data *D*:
a1, a2, a4, a5 ... .
a4, a5, a3, ...
a1, a2, a6, ...
... a4, a5 .

a1, a2, , a5, ... , a10.
a1, a2, a4, , ... , a10.

KeyGraph Step 2) Obtain *roofs* (in FC5) or *bridges* (in FC6), i.e., items co-occurring with multiple bases or islands respectively. If the node is of lower frequency than black nodes (e.g. a10), it is a new node put as a red one. Otherwise it is a black node surrounded by circles (e.g. a4) . And, these nodes represent the candidates of chances in chance discovery.

Fig. 1.7 Two feature concepts of KeyGraphxx

be transformed into a computable model if necessary. In this case, the architecture model originates from a daily requirement for grasping important words despite their low frequency, and is simply put into a computation procedure. On the other hand, when we diverted KeyGraph to chance discovery, that is, to detect and explain an event having a potential importance to human decision-making as in the textile example in the preface, a new feature concept has been proposed, that is, *islands and bridges* [19]. For example, suppose the target is the POS data of supermarket as in D2 above (see Fig. 1.7), the composite items can be as follows:

- An island, which is called the basis in the architecture model, is a cluster that consists of black nodes linked by solid lines. For example, the set of snack items as "chips," "cheese snack," "cracker" forms one island, and liquor items "beer," "wine," "sake" forms another.
- A bridge is defined as the connection of islands, as visualized by dotted lines. A bridge may contain nodes that do not belong to the island, as relay points of dotted lines, represented by white or red nodes such as "caviar" that appear at a lower frequency than the items on islands (due to the high price in the case of caviar).

The islands can be viewed as the underlying contexts common in the target world because they are formed by the set of items co-occurring frequently in the data set. In addition, the bridges represent concepts or contextual flow that connects the information represented by the islands. In the market example here, a customer visiting the supermarket to buy beer and wine may be interested in caviar, but gives up buying it considering the price and moves to the shelf of snacks to buy chips and crackers. In this case, the caviar worked as a bridge that urged the customer from the shelf of liquor to the shelf of something to eat. Here let us review a case presented in [23], where a map obtained by KeyGraph assisted in chance discovery in real business. The dataset was of a "pick-up" sequence in a textile exhibition organized

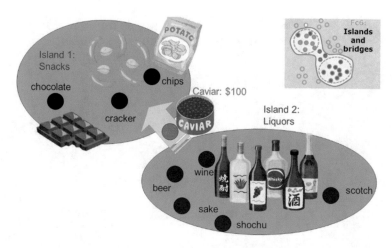

Fig. 1.8 The feature concept islands and bridges (FC6) of KeyGraph

by textile company A, where visitors representing apparel firms or textile-converter firms come to pick up interesting items among more than the 800 exhibited textile samples. Each order card was written manually, showing a set of items picked up by one visitor, and a dataset of as many lines as the number of cards (more than 200) was collected. KeyGraph applied to this dataset showed islands and rare items as bridges connecting the islands. Here, the black nodes linked by black lines indicate islands, and the red parts show bridges (Figs. 1.8 and 1.9).

Ten marketers in this textile firm attached real textile samples to the printed version of the map visualized by KeyGraph, to look at and touch the textile items on it and discuss their inferences about which items will be used, by whom, where, when, why, and how. First, they noticed that two islands correspond to popular item-sets: the large island on the right is an established market of textiles for business clothes, that is, suits, shirts/blouses for under suits, etc. The one on the left, which is just a single node but is an island in the sense that it is frequent, came to be interpreted as textiles for casual, called worn-out, by multiple marketers. A few marketers pointed out that consumers desire to move from one island to another. For example, when a woman moves from the workplace to a restaurant for dinner after working, she may like to change clothes from business suits to casual for relaxation. Interested in such consumers, marketers came to pay attention to the infrequent items between the islands (the new corduroy in the dotted red ellipse in Fig. 1.9). The marketers then obtained a strategic scenario to sell the new corduroy to suit manufacturers as casual jacket office workers can wear easily fitting the trousers of suits to go out for dinner. Similarly, they externalized the value of other infrequent items and reached hits in sales. In this example, the new corduroy showed a chance because it affected the decision of the marketers and expected consumers of the new casual jacket. We can say that the marketers worked as conceptualized in the feature concept of *islands and bridges*.

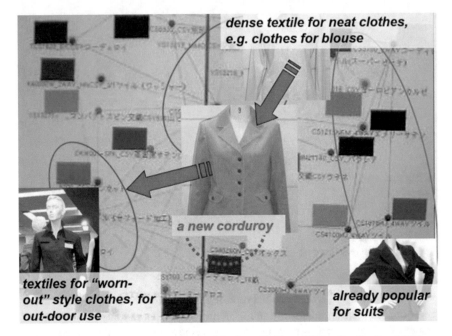

Fig. 1.9 A realization of the feature concept islands and bridges (FC6) using KeyGraph [23]

On the other hand, Example 3 was realized by introducing a feature concept *"explanatory change"*, *"trend shift"*, and *"diversity shift"*. The solution Sol3 was embodied after the presentation of the original solution *"explain changes in the market by showing causal events such as items or behaviors of customers"* via involving *"trend shift"* and *"diversity shift"* as the sub-feature concepts of *"explanatory change"* which was not easy to illustrate as an image.

Example 3 **Change explanation in businesses**
Req3: Detect and explain causal events in the tipping, that is, changing points of consumer behavior in the market
Sol3: Explain changes in the consumption market with visualized "explanatory changes" implying the latent dynamics such as the "trend shift" or "diversity shift" in the market.
DJ3-1: Market data, for example, position of sale (POS) in a supermarket or stock prices
DJ3-2: Data on social events and news
TJ3-1: Tangled String or some other tools for explaining a change.

In this example, a TJ stands for a tool jacket, where a tool for using data (a method of AI, data visualization, or simulation) is summarized in a form similar to DJ, that is, the title (e.g., *KeyGraph*), the abstract (e.g., *visualizing the co-occurrence relations between both frequent and infrequent items in the data*), and the input/output variables (*word, item, event, human, time*, etc.). Tangled string (TS) is a method for explaining a change by positioning an event in a string representing a sequence that tangles on the way if an event occurs multiple times, as shown in Fig. 1.10 [28]. In the sense that there are different time ranges for different trends that can be connected by periods bridging a trend and the next trend, the history in the market can be regarded as a string that has some tangled parts. Here, the feature concept is a *tangled string*, so using TS has been proposed as a way to realize the solution in Example 3, that is, change explanation. Change explanation can be logically represented using the predicate "change" defined indirectly in clause Eq. (1.6), i.e., indirect in the sense that the defined predicate is in the RHS, not LHS. This means to position a change as a transition from a trend in a certain period $(t - \Delta t : t)$ to the next period $(t : t + \Delta t)$ if the market changes substantially, as in Eq. (1.5) and to explain the trends logically.

$$|market(t : t + \Delta t) - market(t - \Delta t : t)| > Q \tag{1.5}$$

$$trend_1(t : t + \Delta t) : -trend_2(t - \Delta t : t), change(t) \tag{1.6}$$

$$market(t : t + \Delta t) = vector_i : -trend_i(t : t + \Delta t) \tag{1.7}$$

The cause of the state of the market is given by the changing trend, as in Eq. (1.6). The requirement is to explain the causality of the change, as in clauses Eqs. (1.6) and (1.7). The difficulty is that the value of only the predicate "market" can be obtained from the data in hand, which is linked to $trend_i$, which is the i-th trend (countable by 1, 2, ... i, ... N, where N represents the number of known trends, e.g., $trend_1$ = hot meal, $trend_2$ = beverage for cooling the body, etc. in the food market), which may not be included in the given data but can be linked to external events. It is emphasized that the "trend" outside of the data may be interpreted using humans' common sense. In this sense, we need to link the data-based finding to external information. It is essential to distinguish the explanation of changes from the detection or prediction that have been realized using machine learning technologies (e.g. [8, 12, 18]). By applying TS, market$(t - \Delta t : t)$ before and market$(t : t + \Delta t)$ after each change point t are visualized as two substrings called pills, that is, the tangled parts, corresponding to a trend where the same items are repeatedly purchased. Here, $trend_2(t - \Delta t : t)$ and $trend_1(t : t + \Delta t)$, which are the interpretations of the latent trends of market$(t - \Delta t : t)$ before t and market$(t : t + \Delta t)$ after t by experts in the markets (based on the visualization if we can apply a tool for it), could be explained by relating their external knowledge, that is, knowledge acquired from daily business but out of the data. TS itself was originally created as a result of IMDJ, where the following requirement, solution, and DJs have been proposed as a set.

Req 0: Collect credible and persuasive information for to explanation
Sol 0 : Extract high impact information
DJ0-1: Log text of communication
DJ0-2: Facts for supporting/rejecting messages

The role of TS in the original version [22] was to detect a message with the highest social impact from a log of human communication (DJ0-1), for realizing the solution Sol 0 to satisfy the requirement Req 0, after which additional information (DJ0-2) should be imported, via human interpretation of the string visualized by TS, to validate the credibility of the message corresponding to the chosen messages. Here, a sequence of words has been modeled by a string, of which each entangled part is called a "pill" where some items or events appear multiple times within a certain length. Pills may meet each other when they share the same items to make a larger pill. The entire sequence can be divided into pills and wires that are substrings connecting pills, corresponding to a complex idea composed of a mixture of simpler ideas. The key items in the sequence are positioned as (a) and (b) below on the tangled string.

(a) Key item in the pill: a repeated item in a pill
(b) Key item on wire: an item which initiates or concludes a pill

See Fig. 1.10a for the summary of the TS algorithm (details are found in Refs. [22, 28]). Figure 1.10b is a photo of the author himself warming up his own idea by tangling a paper string by cutting a printed document before coding the program for the TS algorithm. Such a free-hand action is useful as a part of metacognition for externalizing a feature concept on one's mind. An example from [22] is for the target log text of a debate in the Japanese national parliament in 2004 visualized by using TS, as shown in Fig. 1.11, where the passing of time corresponds to the direction in the right. The white lines here represent tangled clusters, that is, pills, whereas the red (dark colored in B/W) the wires. Letters on the yellow (white in B/W) rectangles represent key items in pills, whereas those on red are key items on wires. The letters in the figure are shown in Japanese, so we put English words in the added letters and briefly itemize noteworthy parts below.

The extent of tangling, that is, the size of pills, at each time in the discourse can be observed as the height of lines—large pills as white lines reach high parts of the figure, red lines also for connecting large pills. For example, a large pill is found just after the word "hostage" in the left of the figure, meaning the enthusiastic

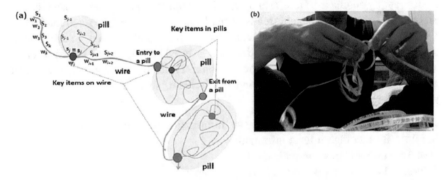

Fig. 1.10 An illustration about the generation of Tangled String from a sequence in **a**. The photo of the author, that happened to be taken before the development of TS in **b**

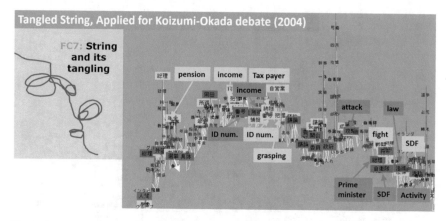

Fig. 1.11 TS applied to the discourse log in a parliament of Japan in 2004, with the corresponding feature concept (reinterpreted the result in [22] by an FC, changing the highlighted words)

discussion triggered by the topic of Japanese hostage captured in Iraq. After this word, we find a period where various topics such as "pension," issues relevant to prime minister such as "election," etc. are mixed. The highest part in the middle of the figure includes various topics such as "welfare," "self-defense force" etc. and are triggered by the accusation of the prime minister for his compulsory decision in the committee of "welfare," as shown by the red key item on the wire (red rectangular nodes). The "self-defense forces (SDF)" of Japan, as well as the tax payers' "ID number" and "hostage" appear as key words on wires. From these results, we found that central concepts that influenced later changes in Japan were extracted by TS, and typical patterns of individual discussants were noticed. Especially, "hostages" here that triggered the reinforcement of long-lasting interest of Japanese people in the activities of SDF, connected to other essential topics for Japanese history. The words in yellow, representing key items in pills, include such words as "tax" and "self-employed" firms, which turned out to be significant social issues that still continues in Japan after increasing the consumption tax rate in 2014.

TS was then applied to Example 3, and as a result, Fig. 1.12 has been obtained for a supermarket in (a) and stock market in (b) ((b): result of a similar analysis to [28]). The encouraging results include two points: Point (1) The connections from/to trends in the market visualized aid in explaining the changes in the entire target market. From (a), supermarket managers pointed out that the high-resolution detection of turning points from customers' interest in preparing for lunch, dinner, and breakfast of the next day is helpful for aiding the shifts or salesclerks, as shown by the arrow in Fig. 1.12a. From (b), the visualized changes in the stock market could be associated with middle-length term changes caused by political decisions and long-term changes due to innovations in the industries corresponding to the stocks. Point (2) The change points found as entrances to pills of TS coincided with high precision with the real increase in the price of the stock appearing at the entrance. The strength of TS here is that the trend shifts in both the entire stock market and the price of each stock,

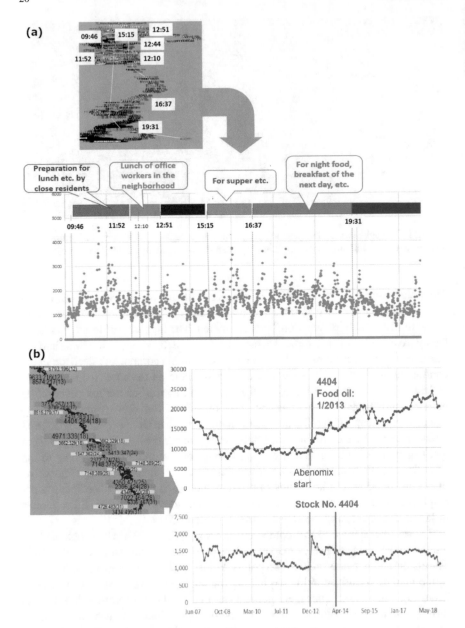

Fig. 1.12 TS applied to the **a** supermarket POS data and **b** stock price data

shown by the upper and lower arrows in Fig. 1.10b, respectively, are obtained and explained with one string, according to real workers in stock analysis. Thus, we can relate the changes in a log in DJ0-1 and DJ3-1 with other data representing social trends in DJ3-2.

On the other hand, *diversity shift*, which is our expression of the idea proposed by Kahn (1995) [17] as an explanation of essential trend shifts in a market, works as a feature concept matching the requirement Req3 and computable from data by concretizing as the change in graph-based entropy (GBE). GBE is an index of the diversity of events computed on their distribution to the subgraphs of a co-occurrence graph, as in Eq. (1.8).

$$Hg = \sum_j p(cluster_j) \log p(cluster_j),$$

$$\text{where} \quad p(cluster_j) = \frac{freq(cluster_j)}{\sum_j freq(cluster_j)} \quad (1.8)$$

Here, $p(x)$ is the proportion of baskets, including item-set, and $freq(cluster_j)$ denotes the frequency of events in the POS data, to which cluster j is the closest among all clusters. An event here means the purchase of an item set in a basket by some consumer, so one basket corresponds to one event. In defining the closest cluster to a certain basket, the measure of closeness is defined by the cosine of two binary vectors, that is, $(\vec{u})(u_1, u_2, \dots u_m)$ for basket u and $(\vec{v})(v_1, v_2, \dots v_m)$ for cluster v. In u_i and v_i, the presence of the i-th item in the market is represented by 1 (absence by 0). A cluster here refers to a group of items connecting the pairs of highest co-occurrences. The change in GBE is computed for detecting signs of structural changes in the data and is informative in explaining the latent dynamics of consumers' preferences. For the data on the position of sale (POS), this change can be regarded as a sign of the appearance, separation, disappearance, or unification of consumers' interests that have been regarded as the essence of diversity shifts. In the case of Fig. 1.13, the bridging edge between the two clusters in the graph is cut in the 10th week, which is interpreted as an independent growth of the lower cluster corresponding to spices for cooking stew. This was surprising because the 10th week in the data was a hot period in August, but the result of Google Trend Search supported this by showing that the interest of Google users in "stew" tends to increase from between 17th and 23th of August in Japan every year. Here is an exemplification that the thought beyond the given data could be triggered by the data visualization reflecting a feature concept (FC8 in Fig. 1.13).

We used the DJ store [11], a search engine of DJs that lists not only DJs including the query but also DJs used in previous IMDJ sessions for the purpose of satisfying a requirement or a solution corresponding to the query word. The entered query was "explain a precursor in the changing of earthquakes," to which "POS data" was shown in the output list of the DJ store because a similar requirement (**Req3**: ... explain ...changing points of...) has been give in the marketing case above. As a result, we noticed the same concept of "diversity shift" as used for the POS data in the example

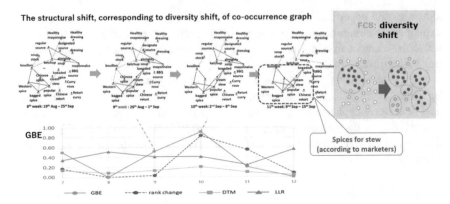

The structural shift, corresponding to diversity shift, of co-occurrence graph

Fig. 1.13 TS applied to the **a** supermarket POS and **b** stock prices: rank change, DTM [4], and LLR [18] are baselines for change point detection

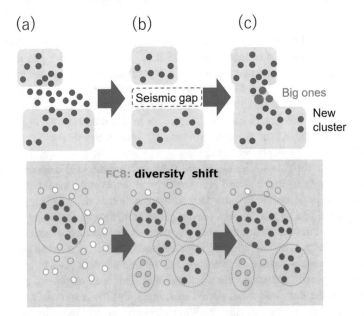

Fig. 1.14 The transition of earthquake activities [21] and its feature concept

above can be used for the analysis of the distribution of epicenters, and invented the following method for satisfying the requirement represented by the above query. Figure 1.14 shows the state transitions during the earthquake activation. In each state from (a) through (e), the appearance of seismic gaps and bridges between clusters of active epicenters are regarded as precursor candidates for earthquake activation. In this case, the author introduced a simple model to explain the precursory process of earthquakes composed of the two phases described below.

Phase 1: The diversity of epicenter clusters increases from state (a) or (b) to state (c) by separating a cluster of active epicenters to create a seismic gap.

Phase 2: The clusters above are combined because earthquakes occur in the seismic gap, as shown in (c). Earthquakes may occur in the seismic gap in the state (e).

The entropy defined on the distribution of epicenters into clusters, called the regional entropy on seismic information (RESI), represented by H in Eq. (1.9), increases in the transition from state (a) or (b) to state (c) and decreases from state (c) to state (d) or (e). Thus, the saturation of the increase in RESI implies a precursory condition for a future earthquake. Thus, here, the diversity shift of epicenters is represented by RESI, borrowing the basic idea of diversity shift in the market represented by GBE. Thus, the common feature concept *diversity shift* or *entropy*, which had been used in marketing, came to be diverted by analogy (base: market dynamics, target: earthquakes) to explain a change that may be regarded as an earthquake precursor. In Eq. (1.9), S is the target region for which RESI is computed. C_i is the i-th cluster of the earthquake foci in S. To capture the dynamics described above, the $H(S, t)$ of region S at time t is expressed in terms of a set of probabilities $p(C_i|S, t)$. The i-th microstate here is interpreted as the i-th cluster C_i of foci in region S to which the focus of each earthquake may belong. A cluster here is a group of closely located epicenters in a period (e.g., one month).

$$H(S, t) = \sum_i p(C_i|S, t) \log p(C_i|S, t) \qquad (1.9)$$

1.4 Discussion: Feature Concepts as Patters for Living Beyond Data

The noteworthy point of the last example in 1.3 is that an FC used for a previous problem can be reused analogically for a new problem with a similar requirement. To understand the contribution of feature concepts to living beyond data, the consideration of similarities and differences from pattern languages [3] is helpful. Feature concepts may be regarded as a customized pattern language, initially proposed by Alexander for urban planning and diverted so far to systems design [5, 9], and to data utility here. In the original version of the pattern language, a set of structural patterns composed of urban elements such as parks, ponds, bridges, houses, etc., were used to explain the existing structures of urban areas and to design a new area. These patterns are supposed to appear typically, that is, having existed in the previous areas as frequently or conceived by people living or working in their respective contexts. In the applications to systems design or organizational activities, each pattern is numbered (1, 2, 3, ...) to which a context, problem (requirement), and a solution to the problem are illustrated for future reuse. The point of being prepared with "typical" patterns is twofold: first, the process to consensus within a team is the necessary condition for going ahead in real businesses, and is smoothened by using

Table 1.1 Comparison of pattern languages and feature concepts (PL has been diverted not only to software design but also for other purposes such as designing organizational activities)

	Patterns in the pattern language on urban design	Patterns in the pattern language diverted to software	Feature Concepts
Goal	Designing or evaluating an urban area	Designing a software	Satisfying requirements in real life
Target system	An urban area as a living environment	Software as a system	Any problem targeted using or beyond data
Pattern: The requirements and the solutions are explained in presenting a pattern	A typical pattern of design corresponding to a part of the area, e.g., school zone, transportation, housing area, park	An abstract representation of subtasks, objects, resources, products, etc., the flow of tasks/information	Each feature concept expected to be obtained from a dataset or federated datasets
Elements	Concrete, mostly known or realistic, entities such as houses, bridges, people, etc.	Abstract, often unknown or imaginary, variables such as X, Y, T, etc., until associated with entities in the real world	Abstract, often unknown or imaginary, variables such as X, Y, T, etc., until associated with data or real world
Preference of pattern	Commonality for abstracting each part of the urban area	Comprehensibility for using, reusing, and sharing with consensus by designers and users	Commonality for reusing, and novelty for creating frequently, a feature concept
Relationships of elements	Hierarchical (combination of vertical and horizontal relations)	Hierarchical, parallel, distributed, and/or sequential	All kinds of relationships, to adopt to the changes in hierarchical, parallel, distributed, and/or sequential structures of real problems and events

the patterns as the common language for communicating the same context, problem, or solution by the same expressions. Second, the patterns should be connected via relationships from/to each other, which may be hierarchical relations or likeliness to be combined. The reuse of a pattern by analogy can be regarded as its connection to a pattern for a previous or a future design. For such a connection, the pattern should have a standardized interface with other patterns or a regulated definition of patterns and their sub-patterns, which may have been implicitly shared in the pattern languages used so far.

Compared with these pattern languages, as in Table 1.1, we can point out that the novelty of each feature concept, created from a novel requirement or novel data, is a unique point. Thus, users of feature concepts are encouraged to be aware of the essential points of feature concepts and be creative enough to create novel feature concepts that may not be accepted soon after the initial proposal. The uniqueness of a feature concept, meaning the difference of a conceptualized feature from others' ideas and of its goal from others due to being originated from a novel requirement of someone in the society can be a priceless treasure for the society—even if the one is just a single individual.

However, once a feature concept is created and shared with others, it is a typical and useful tool for innovators who federate and/or use data, which is similar to a pattern language. In addition, relationships between feature concepts can be, similarly to patterns in a pattern language, the hierarchical structure (e.g., "network of networks" over "network," "diversity shift" over "diversity," etc. where the latter is a part of the former), the connectivity (e.g., connect "diversity shift" and "islands and bridges" for explaining the change in the structure of an organization), or the interactive relation as parts of a process toward data-federative innovation. More importantly, the feature concepts are linked to requirements, as patterns in pattern languages are. These structures and the elements (e.g., "nodes" and "edges") of the structures (e.g., "network") in a feature concept can be created without the consideration of team-wise consensus at first, because a feature concept is created for designing a novel tool (of e.g., machine learning or visualization) for enhancing data utility in the context of competition of developers and data scientists.

Furthermore, the elements and the composed structures of a feature concept are both abstract and conceptual, as implied by the name feature concepts, independent of entities in the real world until links are created to real-world embodied entities. Such a link comes to be built at the time a data scientist embracing the feature concept in his/her mind communicate the individuals having real life requirements in the real world, or at the time the elements are projected to attributes (i.e., variables) of a dataset or a set of datasets. This point may look differerent from Alexander's pattern language, in which elements correspond to real entities [3], but is partially similar to his abstract formalization of real urbans [2]. This point is similar to the pattern language used in a software design where some activities are defined abstractly, that is, independent of the real behaviors of the software or its user(s), but is linked to reality when the target users and software are embodied.

Here, let us remind our basic principle in IMDJ that datasets are created, provided, or opened after the elicitation of requirements or problems and solutions to them, in an innovative communication referring to DJs. Furthermore, the names of feature concepts and their elements are difficult to understand for others than their creators because they may be metaphors rather than concrete entities, due to the nature of data they deal with, which are abstract forms of information that are independent of, although originated from, entities in the real world. Thus, we can say that the links from feature concepts to things, events, and actions in the real world should come from the open-minded communication of data providers and data scientists, with users living in the real world where they declare requirements and ask why and how the requirements are essential.

1.5   Conclusions

Feature concepts have at least two opportunities to contribute to users' living beyond the data. The first is in its use for establishing a common frame for evaluating and improving the obtained pattern. For example, suppose the feature concept composed of nodes and arrows as FC4 above has an empty causal event, that is, a node without an assigned real entity, that is, an item or an event, which is not pointed out by other nodes. This missing cause in the obtained structure may mean a low evaluation by the user, but enhances his awareness that he should discover a causal event from external information. The second is the use of the feature concept seeking its novelty. For example, as in the case of creating the "island and bridge" (FC6), its use meant to highlight an infrequent and uncertain event as a potential resource for businesses although such an event had been discounted in those days' (about year 2000, just before [19]) trend of machine learning to acquire frequent patterns from data. Furthermore, in the progress of studies on chance discovery, we developed methods for externalizing unseen scenarios of actions and events toward the future and relating the results of visualization to external information out of data, and also to extend islands and bridges to crystallization as newer a feature concept [20]. This means seeking a novel dimension in the performance of data mining, corresponding to Drucker's definition of innovation, that is, to shift from seeking a believable (statistically significant) pattern to a new concept, shedding light on new discoveries that could not be found from the previous view points in using data. Sharing and using/reusing feature concepts are not only a technology but also a literacy for innovations that go beyond available data because the elicited feature concepts work as a new dimension itself. In this sense, a useful idea for data-federative innovation may be agile engineering by creating and re/using feature concepts on the way of adding and using data and promptly communicating ideas using the corresponding feature concept, rather than taking time to reach a complete consensus among teammates with full understanding of each other at each level from elements to sub-systems which may not relevant to the essential feature concept representing the common interest of the team.

References

1. Agrawal, R., Srikant, R., et al.: Fast algorithms for mining association rules. In: Proceedings 20th International Conference Very Large Data Bases, VLDB, Citeseer, vol. 1215, pp. 487–499 (1994)
2. Alexander, C.: Notes on the Synthesis of Form, vol. 5. Harvard University Press (1964)
3. Alexander, C.: A Pattern Language: Towns, Buildings, Construction. Oxford University Press (1977)
4. Blei, D.M., Lafferty, J.D.: Dynamic topic models. In: Proceedings of the 23rd International Conference on Machine Learning, pp. 113–120 (2006)
5. Buschmann, F., Meunier, R., Rohnert, H., Sommerlad, P.: Stal, Pattern-oriented Software Architecture. A System of Patterns. Wiley (1996)
6. Drucker, P.F.: The discipline of innovation. Harv. Bus. Rev. **63**(3), 67–72 (1985)

7. Eris, O.: Effective Inquiry for Innovative Engineering Design, vol. 10. Springer Science & Business Media (2004)
8. Fearnhead, P., Liu, Z.: On-line inference for multiple changepoint problems. J. R. Stat. Soc. Ser. B (Statistical Methodology) **69**(4), 589–605 (2007)
9. Foote, B., Rohnert, H., Harrison, N.: Pattern Languages of Program Design 4. Addison-Wesley Longman Publishing Co., Inc (1999)
10. Furao, S., Hasegawa, O.: An incremental network for on-line unsupervised classification and topology learning. Neural Netw. **19**(1), 90–106 (2006)
11. Hayashi, T., Ohsawa, Y.: Data jacket store: structuring knowledge of data utilization and retrieval system. Trans. Jpn. Soc. Artif. Intell. **31**(5):A–G15_1 (2016)
12. Hayashi, Y., Yamanishi, K.: Sequential network change detection with its applications to ad impact relation analysis. Data Min. Knowl. Discov. **29**(1), 137–167 (2015)
13. Ho, T.K.: Random decision forests. In: Proceedings of 3rd International Conference on Document Analysis and Recognition, IEEE, vol. 1, pp. 278–282 (1995)
14. Ho, T.K.: The random subspace method for constructing decision forests. IEEE Trans. Pattern Anal. Mach. Intell. **20**(8), 832–844 (1998)
15. Hong, C.F., Yang, H.F., Lin, M.H., Lin, G.S.: Creative design by bipartite keygraph based interactive evolutionary computation. In: International Conference on Knowledge-Based and Intelligent Information and Engineering Systems. Springer, pp. 46–56 (2006)
16. Hsu, C.l., Wang, L.h., Hong, C.f., Hsu, F.c., Sung, M.y., Tsai, P.h.: The keygraph perspective in arcs motivation model. In: Sixth IEEE International Conference on Advanced Learning Technologies (ICALT'06). IEEE, pp. 970–974 (2006)
17. Kahn, B.E.: Consumer variety-seeking among goods and services: an integrative review. J. Retail. Consum. Serv. **2**(3), 139–148 (1995)
18. Miyaguchi, K., Yamanishi, K.: Online detection of continuous changes in stochastic processes. Int. J. Data Sci. Anal. **3**(3), 213–229 (2017)
19. Ohsawa, Y.: Keygraph: visualized structure among event clusters. In: Chance Discovery. Springer, pp. 262–275 (2003)
20. Ohsawa, Y.: Data crystallization: chance discovery extended for dealing with unobservable events. New Math. Nat. Comput. **1**(03), 373–392 (2005)
21. Ohsawa, Y.: Regional seismic information entropy to detect earthquake activation precursors. Entropy **20**(11), 861 (2018)
22. Ohsawa, Y., Hayashi, T.: Tangled string for sequence visualization as fruit of ideas in innovators marketplace on data jackets. Intell. Decis. Technol. **10**(3), 235–247 (2016)
23. Ohsawa, Y., McBurney, P.: Chance Discovery. Springer (2003)
24. Ohsawa, Y., Nishihara, Y.: In: Meinel, C., Leifer, L. (eds.) Innovators' Marketplace: Using Games to Activate and Train Innovators. Series Understanding Innovation, vol. 3. Springer (2012)
25. Ohsawa, Y., Benson, N.E., Yachida, M.: Keygraph: automatic indexing by co-occurrence graph based on building construction metaphor. In: Proceedings IEEE International Forum on Research and Technology Advances in Digital Libraries-ADL'98-. IEEE, pp. 12–18 (1998)
26. Ohsawa, Y., Kido, H., Hayashi, T., Liu, C.: Data jackets for synthesizing values in the market of data. Procedia Comput. Sci. **22**, 709–716 (2013)
27. Ohsawa, Y., Hayashi, T., Kido, H.: Restructuring incomplete models in innovators marketplace on data jackets. In: Springer Handbook of Model-Based Science. Springer, pp. 1015–1031 (2017)
28. Ohsawa, Y., Hayashi, T., Yoshino, T.: Tangled string for multi-timescale explanation of changes in stock market. Information **10**(3), 118 (2019)
29. Ohsawa, Y., Kondo, S., Hayashi, T.: Data jackets as communicable metadata for potential innovators–toward opening to social contexts. In: International Conference on Intelligent Systems Design and Applications. Springer, pp. 1–13 (2019)
30. Quinlan, J.: Induction of decision trees. Mach. Learn. **1**(1), 81–106 (1986)
31. Quinlan, J.: c4. 5: Programs for Machine Learning. Morgan Kaufmann Publishers (1993)
32. Takemura, K., Hayashi, T., Ohsawa, Y., Aihara, D., Sugawa, A.: Computational coach support using soccer videos and visualization. IEICE-TR **117**(440), 93–98 (2018)

Chapter 2
Innovation for the Real-World Through Knowing Why

Jan Auernhammer and Larry Leifer

Abstract A key challenge in innovation creation is the rapid generation of new, relevant knowledge. This chapter discusses how design practices generate the fundamental knowledge of knowing why, essential for innovation in the real world. It illustrates the importance of knowing why in uncertain and ambiguous situations and the conditions of innovation. We show that design practices, such as need-finding and prototyping generate knowing why. This knowledge is inherent in action and practice and generates a challenge for an observer when researching innovation performance. Knowing why and knowing how are not in the observer's knowledge domain. Therefore, we emphasize that a *productive science* is required to produce the combination of knowing why, knowing how, and knowing that to understand and create innovation.

Keywords Innovation · Design · Knowledge generation · Productive science · Ambiguity

2.1 Introduction

The Design Division/Group at Stanford University developed and cultivated a design practice and pedagogy that focused on creativity, people, and collaboration [5]. We view *design* as the response to and translation of people's needs into a tangible design solution [15]. This perspective has several implications for research. First, unlike the technical and rational worldviews, we do not see artifacts and technology by themselves as an innovation. We hold a humanistic view in which innovation is the enhancement of the living world, including society and nature. Second, this view requires various types of knowledge, particularly the type of.

In this chapter, we discuss the importance of *knowing why* for innovation. We illustrate that *knowing why* is essential in situations of high uncertainty and, the con-

J. Auernhammer (✉) · L. Leifer
Center for Design Research, Stanford University, Building 560, 424 Panama Mall, Stanford, CA 94305, USA
e-mail: jan.auernhammer@stanford.edu

© Springer Nature Switzerland AG 2023
Y. Ohsawa (ed.), *Living Beyond Data*, Intelligent Systems Reference Library 230,
https://doi.org/10.1007/978-3-031-11593-6_2

ditions of innovation. We outline that design practices create *knowing why* through various activities, such as need-finding , and prototyping, and this knowledge drives the creation of innovation. These activities allow grasping (1) human needs and motivation, the *Necessity* of innovation, and developing (2) theory-in-action, the *Capability* of innovation [17]. This chapter addresses the fundamental observer problem of creating knowledge in the case of innovation and outlines the essential practices to generate knowledge for innovation.

2.2 Knowing, Uncertainty, and Innovation

The term *knowing* has its origins in cnāwan (old English), which means "recognizing" or "identifing." "To know" means to operate adequately in individual and cooperative situations [14]. However, innovation results from a situation in which there is no complete or adequate knowledge. The implication of incomplete knowledge in innovative performance and the importance of knowing why are discussed in more detail in this section.

2.2.1 Ways of Knowing in Innovation

Various types of knowledge are essential for innovation. The first type of knowledge is knowing-that. Knowing-that (epistêmê) provides explanations. This type of knowledge is generated by examining and studying existing phenomena. Knowing that incorporates explanations and descriptions of general rules and theories produced by systematic methods. When an individual knows that something is the case, this knowledge can be transmitted through speech and writing. The second type of knowledge is *know-how*. Knowing-how (technê) is the embodied knowledge and craftsmanship of established practices. Knowing how is situated in people's experiences and inherent in their everyday actions and practices and is produced by people reflecting on their actions [2, 18]. Knowing-that and knowing-how are inseparable as, e.g., scientists could not discover any particular actualities (*knowing-that*) unless they know how to discover (*knowing-how*), and designers could not create any particular artifact or system (*knowing-that*) unless they know how to design and build it (*knowing-how*).

However, both types of knowing are insufficient in innovation. Innovation requires knowing before the fact. This lack of knowledge implies that design teams, in the beginning, do not know what to build (knowing-that) and do not know how to create it (knowing-how). It produces a situation of high uncertainty of not knowing the value and ambiguity of not knowing or grasping the whole situational structure. Design teams need to dynamically create *knowing-that*, e.g., creating and naming the new design solution, and they need to create *knowing-how* to build this solution. When dynamically creating knowing-that and knowing-how, the driving force that allows taking action is *knowing-why*.

The third type of knowing is **knowing-why**. Knowing-why is often expressed as Necessity. Plato in the *Republic* expressed that "our need will be the real creator." Knowing-why a need exists allows improving the situation. Knowing-why is also *Capability* is knowing-why a certain action leads to just this kind of effect in situations with similar structural conditions [20]. Schön [18] and Argyris and Schön [2] expressed this as. In the next section, we will illustrate the importance of knowing-why and why it is required to be able to take action in situations of uncertainty and ambiguity.

2.2.2   Knowing why: Knowing in Situations of Uncertainty and Ambiguity

In the beginning, one does not know-how to operate adequately in the situation required for innovation and cannot describe the thing that one is attempting to build. The creative process starts with a hunch, question, or sensing a problem, need, or desire. It is a process that requires productive thinking of creating new knowledge [10].

To illustrate the problem of knowing in uncertain and ambiguous situations, we will use and expand an example introduced by Maturana and Varela [14]. They used this example to illustrate that the genetic and nervous system does not simply code the information about the environment and represent it in their functional operation. We will use this example in a more literal sense of knowing and designing to illustrate the importance of *knowing-why* in situations of high-level uncertainty and ambiguity, the conditions of innovation. Maturana and Varela ([14], p. 53) outlined, "[l]et us suppose that we want to build two houses. For such a purpose, we have two groups of thirteen workers each. We name one of the workers of the first group as the group leader and give him a book which contains all the plans of the house showing, in the standard way the layout of walls, water-pipes, electric connections, windows, etc., plus several views in perspective of the finished house. The workers study the plans and, under the guidance of the leader, construct the house approximating continuously the final stage prescribed by the *description*." In this case, the workers followed a description of the house (*knowing-that*). The knowledge of what a house is within their resulting in the reproduction of existing solutions.

"In the second group, we do not name a leader; we only arrange the workers in a starting line in the field and give each of them a book, the same book for all, containing only neighborhood instructions. These instructions do not contain words such as house, pipes, or windows, nor do they contain drawings or plans of the house to be constructed; they only contain instructions of what a worker should do in different positions and in the different relations in which [s/]he finds [her/]himself as [her/]his position and relations change" ([14], pp. 53–54). In this case, the workers did not know what they are building (*knowing-that*). They only knew how to build it, the instructions (*knowing-how*). Knowledge of an existing solution is encoded in each individual worker's action and practice. The knowledge of how to build a house is within their collective knowledge domain. This system dynamics exist in large global distribution chains in which nobody knows the entire chain of what the end result is as everyone contributes with their local production. They produce an existing system through each individual's know-how.

We are now going to introduce a third case that represents the knowledge challenge in the situation of innovation. The situation in which an innovative performance is accomplished starts with not knowing what to build (knowing-that) and how to build it (knowing-how). In this third group, we do not name a leader, we do not arrange the workers, and we do not give them a book that describes any instructions. The workers have prior knowledge about the world, how to interact within it, and the situation provides necessary resources. Knowing what a house is and how to build it does not exist within their knowledge domain. Therefore, this knowledge needs to be created. In such an ambiguous situation, the workers have to collaboratively figure out what they should build and how to build it. This process will be a back and forth between the workers, figuring out a solution (*knowing-that*) and how to organize everyone in such a way that they are able to take collective action to build what they are attempting to build (*knowing-how*). In such a situation, the driving knowledge is neither know-that nor know-how. For this reason, *knowing-why* becomes essential as it is unlikely, and even counterproductive, to start building anything until the workers *know-why*, the grasping, and understanding of some kind of need and motivation. Additionally, before building anything, the workers need to figure out how to create the structure, including their actions and organization that will result in a feasible solution. It requires developing *knowing-why* this kind of action and organization leads to just this kind of effect and outcome within the given situation, the theory-in-action [2]. It is developing the instructions that result in a new and feasible solution.

In the first and second groups, *knowing-why* is implicitly encoded in the description and instruction of the house. Someone knows why a house is needed, and someone knows why the proposed structure of the house will result in a well-founded house. Knowing-why is not essential for the workers in accomplishing the task of build the house. However, in the last case of high uncertainty and ambiguity, *knowing-why* becomes essential. In this case, there is no instruction and description of the final outcome. The workers have to figure out what and how to build anything in the first place, which is ambiguous process. Therefore, the workers have to figure out why a solution is needed (human need), why are they building this solution (human motivation), and why a certain structure, organization, and collaborative actions will result in a feasible and viable solution (theory-in-action). *Knowing-why* drives the workers' action (necessity) and allows them to take action and to reorganize (capability).

2.2.3 The Observer Problem in Research

These different types of knowing generate an issue for scientific research. In the first and second case, an observer will observe and describe that the workers are building a house (knowing-that). The observer will not observe the difference in the agency of the workers (know-how). In the second case, the workers do not know that they are building a house as it is not in their cognitive domain. It is only in the domain of the observer. In the third case, the observer will probably witness several attempts of building various versions and will recognize the solution when it is finished. From the perspective of observers, the need, desire, and motivation of the worker and why the structure of the final solution is producing this result are invisible as the knowing-why is only in the cognitive domain of the workers. For this reason, knowing in the process of innovation is only in the knowledge domain of the workers and not the observer. The observer is able to generate knowledge after the fact by examining the innovation outcome (knowing-that). The observer is also not able to observe the difference in operation and practice (knowing-how) from her/his point of view. There is no difference between the first and the second case. The third case will look like a very chaotic process in which the observer can make sense of when the solution is accomplished.

Furthermore, inductive and deductive inference based on the observation results in abstractions and generalizations. These abstractions are fragmentary pieces of two or many objects (houses and their construction) or comparison regarding just any features in two or many objects resulting in context-independent similarities [20]. This approach and its results ignore context-dependent situational structure, particularly the know-how and know-why. Generating knowing-why, the need, motivation, and theory-in-action is situational and requires productive thinking or inference [12, 20]. For this reason, innovation research requires a *productive science*, which we call *Design*. It is the practice that iteratively creates *knowing-that* and *knowing-how* on the basis of *knowing-why*.

2.3 Knowing why: Necessity and Capability

Knowing-why is the grasping and knowing of *necessity*, people's needs and motivation and developing *capability*, the action-outcome relationships, or the theory-in-action [5, 17]. Both are essential in the process of creating innovation, the creation of an invention that is needed and valued by people and therefore embraced and obtained. This section will describe that *knowing-why* is comprehended in the interrelation between individuals (the knowers) and their environment (context).

2.3.1 Necessity: Human Needs and Motivation

The first knowing-why is a *Necessity*. Necessity, the human needs, is the underlying driver of people's perception, action, and behavior [7, 8, 21]. Grasping people's needs allows *knowing-why* people act in a certain way, find meaning in their activities and lives, value a specific solution, and designers are intrinsically motivated in creating this design solution over another. Grasping and knowing this driving force in people's lives and inherent in design activities provide the opportunity to create valuable and meaningful solutions for people and is essential in producing a tangible outcome.

Grasping and knowing a human need requires understanding people in relation to their cultural and natural context, as needs do not exist in isolation. They only exist in situational context as needs are produced in the interaction between the individual and the cultural and natural environment [15]. A human need arises when the cultural and natural environment produces a situation in which individuals are not able or unaware of the situation to respond adequately. For example, the workers in the third case can explore what is needed within their community and what they should build to provide value. If the community requires shelter, a house might be appropriate. If there is a lack of drinking water, they may use their resources to design and build a solution that captures rainwater. Knowing-why (a need) determines the generation of knowing-that (a solution) and knowing-how (activities and organization). The identified and felt needs by the workers become drivers for invention. From this perspective, *Design* is the response to a human need [15].

2.3.2 Capability: Theory-in-Action

The second *knowing-why* is capability. Capability is knowing-why just this kind of action leads to just this effect for intrinsic structural reasons [20]. Grasping why this action leads to this effect and outcome allows individuals to vary their actions in a structurally sensible way when the situation is no longer the same, the theory-in-action. Such knowledge is inherent in professional practice [18].

This *knowing-why* is essential in design to produce a tangible and feasible solution. Many solutions on paper, e.g., a design sketch, does not work in reality due to their structural incompleteness or mismatch. Discovering, grasping, and understanding why a designed structure works in specific situations is part of the innovation task. A well-known example is the prototype of the Forth Bridge located in Scotland, as shown in Fig. 2.1.

The prototype illustrates the underlying principle of why this kind of structure produces a tangible solution. The bridge, still in operation, was completed in 1890 and is a UNESCO World Heritage Site. In our example, the workers have to develop *knowing-why* any structure will result in a feasible solution and organize their actions (instructions) to build this structure. They have to dynamically develop the capability of knowing-why this kind of action and organization leads to the desired

Fig. 2.1 Forth Bridge Prototype to explain the principle (theory-in-action) behind a cantilever bridge

outcome. Evaluating prototypes validates knowing-why. Knowing-why (theory-in-action) determines knowing-how (instruction of organization) and knowing-that (tangible solution).

Necessity and Capability are essentially about knowing-why in the creation of innovative design solutions. Grasping people's needs and motivation (knowing-why) requires observing and engaging people in their interrelation with their environment. Developing the theory-in-action (knowing-why) requires prototyping and evaluating a design solution for its structural purpose. Design practices including Need-finding, Benchmarking, Prototyping, and Testing generate *knowing-why*.

2.4 Design Practices: Knowing why

Since the 1950s, the Design Division/Group at Stanford University developed a design practice and pedagogy to iteratively create *knowing-why* [1, 3, 16]. The first design practice that creates knowing-why is *Need-finding and Benchmarking*. Creating *knowing-why* by comprehending a need triggers new ideas and solutions through *Imagination*. The second design practice that discovers theory-in-action is *Prototyping and Evaluating*. Creating *knowing-why* by comprehending an action-outcome relationship allows developing a tangible solution. These design practices are a strategy of iteratively developing knowing-why to drive knowing-that, the new idea or solution, and drive knowing-how, the organization of action towards a tangible solution. These activities form a design cycle, as illustrated in Fig. 2.2.

Fig. 2.2 Design cycle that
creates knowing-why:
necessity and capability

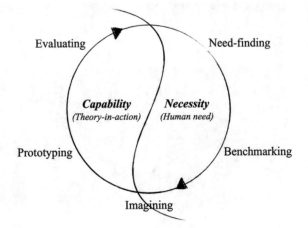

2.4.1 Necessity: Need-Finding and Benchmarking

Robert McKim developed the *Need-finding* practices in the 1960s. Need-finding is
a set of activities to examine people and their life situations, including historical,
social, cultural, artificial, and physical context, to identify "situational tensions." Situational tensions occur when the environment produces a situation in which "something is missing," and people are not able to change this situation actively. These
situations result in human needs, such as physiological, safety, love and belonging,
esteem, self-actualization needs, and a combination of them [13]. Grasping these
situational tensions results in *knowing-why* a situation produces a need and provides
an opportunity to be enriched for people. For example, the workers identify that it
takes a lot of effort to bring water from the nearby lake to the fields by examining
why field workers are exhausted and why areas of the field are running dry, leading
them to design a solution for this need. Another more current example is examining
why a young teenager is depressed (tension) when spending time on social media
platforms. Situations in which the young teenager is comparing her/himself with
unrealistic representations of others due to photo filters result in a lack of self-worth,
which produces this tension in her/his life. Grasping this tension within situational
context provides knowing-why a new solution is needed triggering the question: How
might we design a healthier life situation for the teenager?

Benchmarking activities are part of generating the knowing-why. The identification of something is necessary. Grasping why current design solutions do not produce
a healthy, meaningful, or valuable situation in people's lives provides clues of why
this tension in situational context exists. It is an experiential and analytical activity.
These activities discover the inherent problems of existing solutions and provide a
particular perspective of what is needed to solve them, triggering new ideas (imagining). Duncker [10] describes several analysis techniques to evaluate solutions, such
as conflict analysis to reveal what is really needed to solve the design problems.
Knowing-why is an essential part of the creative process. It is identifying a pro-

ductive question. Wertheimer ([20], p. 123) emphasizes that "[...] the function of [productive] thinking is not just solving an actual problem but discovering, envisioning, going into deeper questions. Often in great discovery, the most important thing is that a certain question is found. Envisaging, putting the productive question is often more important, often a greater achievement than a solution of a set question [...]." Need-finding and Benchmarking discover the more profound question that triggers a productive thought process towards new solutions. These practices are essentially about recognizing and identifying *necessity*.

2.4.2   Capability: Prototyping and Evaluating

After envisioning a new solution that potentially changes the situation in which the need exists, the second type of knowing-why, the theory-in-action, is essential. Theory-in-action is developed through *Prototyping and Evaluating*. Prototyping is about creating a structure representing critical aspects of the solution. Evaluating this structure uncovers why this action and structure leads to the desired outcome or not [19]. Prototyping and evaluating is an iterative process of creating knowing-why the solution is solving the tension, why certain material structures produce a feasible solution, or why certain organization results in the production of the solution. In this process, learning from dead-ends and failures is essential as it allows reframing the problem in a more productive way to notice the theory-in-action [9]. Prototyping and Evaluating essentially create knowing-why a design solution works or not within the specific situational context. In our example, the workers have to figure out which structure in subsequentially which collaborative action and organization will result in producing a solution that changes the situational tension. Prototyping and Evaluating are essentially about developing the *capability* to generate a tangible outcome.

The activities of Need-finding, Benchmarking, Imagining, Prototyping, and Evaluating are an iterative cycle, a back and forth until a necessary solution (knowing-that) and the capability to be able to build (knowing-how) a tangible outcome is accomplished. Knowing-why is the knowledge that drives action and solutions in uncertain and ambiguous situations.

2.5   Conclusion

Design, as developed and cultivated at Stanford University, is the practice that creates knowing-why, which drives the knowing-that and knowing-how in uncertain and ambiguous situations, the condition for innovation. These practices generate knowledge outside the observer's knowledge domain. It can only be accomplished by doing. For this reason, problem and project-based learning is part of the design pedagogy [11]. This practice that generates context-specific knowledge is based on productive thinking generate to context-specific knowledge [12, 20]. These practices are utilized in the design of products, organizations, and culture to facilitate innovation [4, 6]. Design is a *productive science* that produces knowing-why (needs and theory-in-action), knowing-how (actions and organization), and knowing-that (designed situations). *Design* is the practice that creates the knowing that results in innovation for the real-world.

References

1. Adams, J.L.: Conceptual blockbusting: a guide to better ideas. Basic Books (2019)
2. Argyris, C., Schon, D.A.: Theory in Practice: Increasing Professional Effectiveness. Wiley (1992)
3. Arnold, J.E.: Creative Engineering Seminar (1959)
4. Auernhammer, J.: Design research in innovation management: a pragmatic and human-centered approach. R&D Manag. **50**(3), 412–428 (2020)
5. Auernhammer, J., Roth, B.: The origin and evolution of stanford university's design thinking: from product design to design thinking in innovation management. J. Product Innov. Manag. (2021)
6. Auernhammer, J.M.K., Leifer, L.: Is organizational design a human-centered design practice? In: Proceedings of the Design Society: International Conference on Engineering Design, pp. 1205–1214. Cambridge University Press (2019)
7. Bruner, J.S., Goodman, C.C.: Value and need as organizing factors in perception. J. Abnorm. Soc. Psychol. **42**(1), 33 (1947)
8. Deci, E.L., Ryan, R.M.: The "what" and "why" of goal pursuits: human needs and the self-determination of behavior. Psychol. Inquiry **11**(4), 227–268 (2000)
9. Duncker, K.: Zur Psychologie des produktiven Denkens [The psychology of productive thought]. Springer (1935)
10. Duncker, K., Lees, L.S.: On problem-solving. Psychol. Monogr. **58**(5), i (1945)
11. Leifer, L.: Design-team performance: metrics and the impact of technology. In: Evaluating corporate training: models and issues, pp. 297–319. Springer (1998)
12. March, L.: The logic of design and the question of value. The Architecture of Form (1976)
13. Maslow, A.H.: Motivation and Personality Harper and Row, New York, NY (1954)
14. Maturana, H.R., Varela, F.J.: Autopoiesis and Cognition: The Realization of the Living, vol. 42 (1980)
15. Mckim, R.H.: Designing for the whole man. Creative Engineering Seminar (1959)
16. McKim, R.H.: Experiences in visual thinking (1972)
17. McKim, R.H.: Innovation/Bob Mckim, Charles Bernstein, Janet Anapole. In: Conference on Entrepreneurship at Stanford University [sound recording]: a path to the future (1982)
18. Schön, D.A.: The reflective practitioner: How professionals think in action. Basic Books (1983)
19. Schön, D.A.: Designing as reflective conversation with the materials of a design situation. Knowl.-Based Syst. **5**(1), 3–14 (1992)
20. Wertheimer, M.: Productive Thinking. Harper (1945)
21. Wise, R.A.: Sensorimotor modulation and the variable action pattern (VAP): toward a noncircular definition of drive and motivation. Psychobiology **15**(1), 7–20 (1987)

Chapter 3
Why Abductive Cognition Goes Beyond Just Learning from Data

Lorenzo Magnani

Abstract Abductive cognition refers to all kinds of reasoning to more or less deep hypotheses starting from data, both in humans and in AI machines, which are extremely important in everyday reasoning but also, for example, in scientific practice. In this chapter attention will be focused on those particular kinds of abductive cognition that generate deep results, able to go beyond data thanks to a rich explanatory strength, in which asking "why" is the key-factor. Hypothetical cognition beyond data will be also described thanks to the illustration of *model-based* aspects of abduction, in which the activity of grasping the world, beyond what mere rough data suggests, takes advantage of models, diagrams, and simulations of various nature. Moreover, I will delineate the concept of *manipulative* abduction, first introduced in my book of *Abduction, Reason, and Science*, by showing how we can find methods of manipulative constructivity, in which cognitive processes are centered on the exploitation of external representations to the aim of reaching deep and very abstract results, able to powerfully transcend data, at the same time to create cognitive values "in" and "out" of data. In two separated sections I will illustrate (1) how the excessive emphasis on the merits of the computational programs which learn from data renders human abductive "creative" cognition increasingly assaulted and jeopardized, and (2) the challenges against human abduction and epistemic rigor on the part of what I call computational invasive "subcultures" based on rough computational explanations. Finally, thanks to the analysis of the computational system called *AlphaGo*, I will stress the attention to what I call *locked* and *unlocked* strategies that shed further light on the cognitive aspects of both humans and deep learning machines. The character and the role of these cognitive strategies is strictly related to the generation of cognitive outputs, which range from weak (closer to data) to strong level (beyond data) of knowledge creativity. Locked abductive strategies, even if cognitively weak, characterize (sometimes amazing) kinds of *hypothetical creative*

L. Magnani (✉)
Department of Humanities, Philosophy Section, and Computational Philosophy Laboratory, University of Pavia, Pavia, Italy
e-mail: lmagnani@unipv.it

© Springer Nature Switzerland AG 2023
Y. Ohsawa (ed.), *Living Beyond Data*, Intelligent Systems Reference Library 230,
https://doi.org/10.1007/978-3-031-11593-6_3

reasoning and are typical of intelligent machines based on a massive exploitation of data, but they are limited in *eco-cognitive openness*, which instead qualifies human cognizers who are performing higher kinds of abductive creative reasoning, where cognitive strategies are instead unlocked and are able to transcend data.

Keywords Abductive cognition · Creativity · Data · Locked and unlocked strategies · Deep learning · AlphaGo · AlphaZero

3.1 Abductive Cognition Generates Knowledge Starting from Data

3.1.1 Abduction as Ignorance Preservation

Abductive reasoning always starts from initial data of various kinds, from which appropriate more or less deep hypotheses are generated. A first problem of abductive cognition we have to afford is related to the general problem of the role of ignorance. When we reach a hypothesis thanks to abduction, do we get true knowledge[1] or simply something just provisional that leaves our ignorance more or less intact? As I have illustrated in my second book on abductive cognition [41, Chapter two], following Gabbay and Woods' contention (illustrated in their GW-schema), it is clear that "[…] abduction, that is reasoning to hypotheses, is a procedure in which something lacks epistemic virtue is accepted because it has virtue of another kind" [20, p. 62]. For example: "Let *S* be the standard that you are not able to meet (e.g., that of mathematical proof). It is possible that there is a lesser epistemic standard *S'* (e.g., having reason to believe) that you do meet" [62, p. 370]. Focusing attention on this cognitive aspect of abduction, and adopting a logical framework centered on practical agents, [20] contend that abduction (basically seen as a *scant-resource* strategy, which proceeds in absence of knowledge) presents an *ignorance-preserving* (or, better, an *ignorance mitigating*) character. Of course "[…] it is not at all necessary, or frequent, that the abducer be wholly in the dark, that his ignorance be total. It needs not be the case, and typically isn't, that the abducer's choice of a hypothesis is a blind guess, or that nothing positive can be said of it beyond the role it plays in the subjunctive attainment of the abducer's original target (although sometimes this is precisely so)" [62, p. 249]. In this perspective, abductive reasoning is a *response* to an ignorance-problem: one has an ignorance-problem when one has a cognitive target that cannot be attained on the basis of what one currently knows.

We said that abductive reasoning is a *response* to an ignorance-problem and that one has an ignorance-problem when one has a cognitive target that cannot be attained on the basis of what one currently knows. Typically ignorance problems trigger one

---

[1] In this case truth can be considered as referred to knowledge contents of various kinds, from science to common sense, from the truths about human condition provided by a literary fiction to the ones embedded in the moral rules belonging to a collectivity.

or other of three responses. In the first case, one overcomes one's ignorance by attaining some additional knowledge (subduance). In the second instance, one yields to one's ignorance (at least for the time being) (surrender). In the third instance, one abduces [62, Chapter eleven] and so has some positive basis for new action even if in the presence of the constitutive ignorance.

In abductive cognition the problem is that the hypothesis H is, indeed, *only hypothesized*, so that the truth is not assured. That is, having hypothesized that H, the agent just "presumes" that his target is now attained. Given the fact that presumptive attainment is not attainment, the agent's abduction must be considered as preserving the ignorance that already gave rise to her (or its, in the case for example of a machine) initial ignorance-problem. Accordingly, abduction does not have to be considered the "solution" of an ignorance problem, but rather a response to it, in which the agent reaches presumptive attainment rather than actual attainment. H is just a worthy object of conjecture. It is important to note that in order to solve a problem it is not necessary that an agent actually conjectures a hypothesis, but it is necessary that she states that the hypothesis is *worthy of conjecture*.

Briefly, considering H justified to conjecture is not equivalent to considering it justified to accept/activate it and eventually to send H to experimental trial to get its confirmation or falsification. We have just adopted the *decision* to release H for further work in the domain of enquiry in which the original ignorance-problem arose, that is the activation of H as a positive *cognitive* basis for action. We have to remember that this process of evaluation and so of activation of the hypothesis, is not abductive, but inductive, as Peirce contended. In the framework of the GW-schema it cannot be said that testability is intrinsic to abduction, such as it is instead maintained in the case of some passages of Peirce's writings. This activity of testing, I repeat, which in turn involves degrees of risk proportioned to the strength of the conjecture, is strictly cognitive/epistemic and inductive in itself, an experimental test, and it is an intermediate step to release the abduced hypothesis for inferential work in the domain of enquiry within which the ignorance-problem arose in the first place.

In sum, in the perspective described in this section, through abduction the basic ignorance—that does not have to be considered total "ignorance"—is neither solved nor left intact: it is an ignorance-preserving accommodation of the problem at hand, which "mitigates" the initial cognitive "irritation" (Peirce says "the irritation of doubt"). As I have already stressed, further action can be triggered—in a defeasible way—either to find further abductions or to "solve" the ignorance problem, possibly leading to what the "received view" has called the *inference to the best explanation* (IBE).

It is clear that in the framework of the Gabbay/Woods schema the inference to the best explanation—if considered as a truth conferring achievement justified by the empirical approval—cannot be a case of abduction, because abductive inference is constitutively ignorance-preserving. In this perspective the inference to the best explanation involves the generalizing and evaluating role of *induction*. Of course it can be said that the requests of originary thinking are related to the depth of the abducer's ignorance and to the capacity to transcend data.

3.1.2 Directly Reaching Truth Through Abduction

At the beginning of Sect. 3.1.1 above I have contended that abduction always starts from initial data of various kinds, from which appropriate more or less deep hypotheses are generated: a problem that has to be urgently addressed regards the kind of "distance" of the reached hypothesis with respect to the data from which it originated. I also anticipated that abduction can be either ignorance preserving or truth conferring: in both cases the distance from data can be explained, to a first approximation, in terms of the degree of "creativity" involved, as it will become patent in the following sections of this chapter.

We know that consistency and minimality constraints were emphasized in the classical view of abduction established by many classical logical accounts, more oriented to illustrate *selective* abduction [40]—for example in diagnostic reasoning, where abduction is merely seen as an activity of "selecting" from an encyclopedia of pre-stored hypotheses—rather than to analyze *creative* abduction (abduction that generates new hypotheses).[2] In few words, this perspective teaches us that good abductive hypotheses have to be consistent and plausible (with respect to the background knowledge).

We have to immediately say that these classical requirements (consistency and minimality) certainly do not characterize the deeper cognitive and creative aspects of abduction. in which hypotheses are more distant from initial data and the background knowledge in which they are embedded. For example, to stress the puzzling status of the consistency requirement, it is here sufficient to note that Paul Feyerabend, in *Against Method* [17], correctly attributes a great importance to the role of contradiction in generating hypotheses, and so implicity celebrates the value of creative abductive cognition. Speaking of induction and not of abduction (this concept was relatively unknown at the level of the international philosophical community at that time), he establishes a new "counterrule". This is the opposite of the neopositivistic one that it is "experience" (or "experimental results") which constitutes the most important part of our scientific empirical theories, a rule that formed the core of the so-called "received view" in philosophy of science (where inductive generalization, confirmation, and corroboration play a central role). The counterrule "[...] advises us to introduce and elaborate hypotheses which are inconsistent with well-established theories and/or well-established facts. It advises us to proceed counterinductively" [17, p. 20]. Counterinduction is seen more reasonable than induction, because appropriate to the needs of creative reasoning in science: "[...] we need a dream-world in order to discover the features of the real world we think we inhabit" (p. 29). We know that counterinduction, that is the act of introducing, inventing, and generating new inconsistencies and anomalies, together with new points of view incommensurable with the old ones, is congruous with the aim of inventing "alternatives" (Feyerabend

---

[2] I have proposed the dichotomic distinction between selective and creative abduction in [40].

contends that "proliferation of theories is beneficial for science"), and very important in all kinds of creative reasoning.[3]

Moreover, since for many abductive problems there are—usually—many guessed hypotheses, the abducer needs reduce this space to one. This means that the abducer has to produce the best choice among the members of the available group. Here we touch the core of the ambiguity of the ignorance-preserving character of abduction described in the previous subsection. Why?

- Because the cognitive processes of generation (fill-up) and of selection (cutdown) can both be sufficient—even in absence of the standard inductive evaluation phase—to *activate* and accept an abductive hypothesis, and so to reach cognitive results relevant to the context (often endowed with a knowledge-enhancing outcome, as I have already illustrated in detail in [42]). In these cases the *instrumental* aspects (which simply enable one's target to be hit) often favor both abductive generation and abductive choice, and they are not necessarily intertwined with plausibilistic concerns, such as consistency and minimality.

In general we cannot be sure that our guessed hypotheses are plausible (even if we know that looking for plausibility is a human good and wise heuristic in many cognitive contexts), an implausible hypothesis can later on result plausible. Moreover, when a hypothesis solves the problem at hand, this is enough as to count as solution of the abductive problem (even if, not necessarily a *good* solution or the *best* solution).

In these special cases the best choice is immediately reached without the help of an experimental trial (which fundamentally characterizes the received view of abduction in terms of the so-called "inference to the best explanation"). Not only, we have to strongly note that the generation process alone can suffice, like it is demonstrated by the case of human *perception*, where the hypothesis generated is immediate and unique. Indeed, perception is considered by Peirce, as an "abductive" fast and uncontrolled (and so automatic) knowledge-production procedure. In fact, for Peirce "[...] the perceptual judgments, are to be regarded as an extreme case of abductive inference" [57, 5.181].

Finally, two important consequences concerning the meaning of the word *ignorance* in this context have to be illustrated:

1. abduction, also when intended as an inference to the best explanation in the "classical" sense I have indicated above, is always *ignorance-preserving* because abduction represents a kind of reasoning that is constitutively provisional, and you can withdraw previous abductive results (even if empirically confirmed, that is appropriately considered "best explanations"), in presence of new information. From the logical point of view this means that abduction represents a kind of nonmonotonic reasoning, and in this perspective we can even say that abduction interprets the "spirit" of modern science, where truths are never stable and absolute. Peirce also emphasized the "marvelous self-correcting property of reason" in

---

[3] A rich treatment of the basic "paraconsistent" logical perspectives concerning abduction and the role of inconsistencies is contained in [8].

general [57, 5.579]. So to say, abduction incarnates the human perennial search of new truths and the human Socratic awareness of a basic ignorance which can only be attenuated/mitigated. In sum, in this perspective abduction always preserves ignorance because it reminds us we can reach truths that can always be withdrawn; ignorance removal is at the same time constitutively related to ignorance regaining;

2. even if ignorance is preserved in the sense I have just indicated, which coincides with the spirit of modern science, abduction is also knowledge-enhancing because new truths can be and "are" discovered which *are not necessarily best explanations intended as hypotheses which are empirically tested.*

3.2   How Model-Based and Manipulative Abduction Creates Values "In" and "Out" of Data: The Eco-Cognitive Model (EC-Model)

The perspective illustrated in Sect. 3.1.1, in terms of the ignorance-preserving character of abduction described by the GW-model, is mainly referred to the basic sentential aspects of human cognition, related to the exploitation of natural and artificial languages. However, we have to immediately add that sentential aspects are not canonical for abduction. GW-model perspective does a good job in modeling the ignorance-preserving character of abduction, but makes little contribution to the fill-up problem, that is to the analysis of the mechanisms of generation of hypotheses. I also quickly indicated that it is hard to encompass in the GW-schema cases of abductive cognition such as perception and we can also add, just to make a further example, the generation of models in scientific discovery. A more extended eco-cognitive perspective is needed, also able to show how the cutdown and fill-up problems in abductive cognition appear to be stunningly *contextual*.

Indeed, to the aim of detecting the processes that create value not only "In" but also "Out" of data, as anticipated in the title of this section, we need introduce to the so-called eco-cognitive of abduction (EC-Model), I recently proposed in detail in [43, 45], even if the backbone of this approach can be found in the general analysis of abductive cognition I have provided in [41]. At the center of my perspective on cognition and consequently of abduction is the emphasis on the "practical agent", of the individual agent operating "on the ground", that is, in the circumstances of real life. In all its contexts, from the most abstractly logical and mathematical to the most roughly empirical, I always emphasize the cognitive nature of abduction. Reasoning is something performed by cognitive systems. At a certain level of abstraction and as a first approximation, a cognitive system is a triple (A, T, R), in which A is an *agent*, T is a *cognitive target* of the agent, and R relates to the *cognitive resources* on which the agent can count in the course of trying to meet the target-information, time and computational capacity, to name the three most important. My agents are also *embodied distributed cognitive systems*: cognition is embodied and the interactions between brains, bodies, and external environment are its central aspects. Cognition is occurring taking advantage of a constant exchange of information in a complex distributed system that crosses the boundary between humans, artifacts, and the surrounding environment, where also instinctual and unconscious abilities play an important role. This interplay is especially manifest and clear in various aspects of abductive cognition, that is in reasoning to hypotheses.

My perspective adopts the wide Peircean philosophical framework, which approaches "inference" *semiotically* (and not simply "*logically*"): Peirce distinctly says that all inference is a form of sign activity, where the word sign includes "feeling, image, conception, and other representation" [57, 5.283]. It is clear that this semiotic view is considerably compatible with my view of cognitive systems as embodied and distributed systems. It is in this perspective that we can fully appreciate the role of abductive cognition, which not only refers to propositional aspects but it is also performed in a framework of distributed cognition, in which also models, artifacts, internal and external representations, manipulations play an important role. Already in a passage concerning abduction (that is ἀπαγωγή—"leading away") of chapter B25 of the Aristotelian *Prior Analytics*, we can see some of the current well-known distinctive characters of abductive cognition, which are in tune with the EC-Model. Aristotle is already pointing to the fundamental inferential role in reasoning of those externalities that substantiate the process of "leading away" (ἀπαγωγή).[4] Is We can gain a new positive perspective about the "constitutive" eco-cognitive character of abduction, just thanks to Aristotle himself.[5]

In a wide eco-cognitive perspective the cutdown and fill-up problems in abductive cognition appear to be spectacularly *contextual*. It might seem awkward to speak of "abduction of a hypothesis in literature," but one of the fascinating aspects of abduction is that not only it can warrant for scientific discovery, but for other kinds of creativity as well. We must not necessarily see abduction as a *problem solving device* that sets off in response to a cognitive irritation/doubt: conversely, it could be supposed that esthetic abductions (referring to creativity in art, literature, music, etc.) arise in response to some kind of esthetic irritation that the author (sometimes a *genius*) perceives in herself or in the public. Furthermore, not only esthetic abductions are free from empirical constraints in order to become the "best" choice: as I am showing throughout this chapter, many forms of abductive hypotheses in traditionally-perceived-as-rational domains (such as the setting of initial conditions, or axioms, in physics or mathematics) are relatively free from the need of an empirical assessment. The same could be said of moral judgment: they are eco-cognitive abductions, inferred upon a range of internal and external cues and, as soon as the judgment hypothesis has been abduced and accepted, it immediately becomes prescriptive and

---

[4] The Aristotelian word ἀπαγωγή, in ancient Greek, is indeed advantageously often translated as "leading away".

[5] I have illustrated in detail the various aspects, philosophical, logical, and cognitive of the Aristotelian ideas concerning abduction in [43, 45, 47] and in the recent [50].

"true," informing the agent's behavior as such.[6] Peirce implicity provides various justifications of the knowledge enhancing role of abduction, that is when abduction is not considered an inference to the best explanation in the classical sense of the expression, that is an inference necessarily characterized by an empirical evaluation phase, or inductive phase.

I lack the space to give this issue appropriate explanation but it suffices for the purpose of this study to remember that, for example, one thing is to abduce a model or a concept at the various levels of scientific cognitive activities, where the aim of reaching rational knowledge dominates, another thing is to abduce, as I have already said, a hypothesis in literature (a fictional character for example), or in moral reasoning (the adoption/acceptance of a hypothetical judgment as a trigger for moral actions). However, in all these cases abductive hypotheses, which are evidentially inert, are accepted and activated as a basis for action, even if of different kind.

Assessing that there is a common ground in all of these works of what could be broadly defined as "creativity" does not imply that all of these forms of selective or creative abduction[7] with their related cognitive strategies are the same, contrarily it should spark the need for firm and sensible categorization: otherwise it would be like saying that to construct a doll, a machine-gun and a nuclear reactor are all the same thing because we use our hands in order to do so!

However, a special and rich exploitation of wide eco-cognitive aspects is mandatory to create value "Out" of data such as it is happening in the case of what I called "manipulative" abduction. From an eco-cognitive point of view, in more hybrid and multimodal (cf. [41, Chapter four]) (not merely inner) abductive processes, such as in the case of *manipulative abduction*, the assessment/acceptance of a hypothesis is reached—and constrained—taking advantage of the gradual acquisition of consecutive external information with respect to future interrogation and control, and not necessarily thanks to a final and actual experimental test, in the classical sense of empirical science.

The concept of *manipulative abduction*—which also takes into account the external dimension of abductive reasoning in an eco-cognitive perspective—surely captures, to make a cardinal example, a large part of scientific thinking where the role of action and of external models (for example diagrams) and devices is central, and where the features of this action are implicit and hard to be elicited. Action can provide otherwise unavailable information that enables the agent to solve problems by starting and by performing a suitable abductive process of generation and/or selection of hypotheses. Manipulative abduction happens when we are thinking through doing and not only, in a pragmatic sense, about doing (cf. [41, Chapter one]).

---

[6] I have illustrated above (cf. Sect. 3.1.2) these kinds of abduction, called "knowledge enhancing", that do not need empirical confirmation.

[7] For example, selective abduction is active in diagnostic reasoning, where it is merely seen as an activity of "selecting" from an encyclopedia of pre-stored hypotheses; creative abduction instead refers to the building of new hypotheses. As I have already said above I proposed the dichotomic distinction between selective and creative abduction in [40]. A recent and clear analysis of this dichotomy and of other classifications emphasizing different aspects of abduction I have described is given in [56].

In summary, at least four kinds of actions can be involved in the manipulative abductive processes (and we would have to also take into account the "motoric" aspect (*i*) of inner "thoughts" too). In the eco-cognitive interplay of abduction the cognitive agent further triggers internal *thoughts* "while" modifying the environment and so (*ii*) acting on it (thinking through doing). In this case the "motor actions" directed to the environment have to be intended as part and parcel of the whole embodied abductive inference, and so have to be distinguished from the *final* (*iii*) "actions" as a possible consequence of the reached abductive result. A process that characterizes strong effects both "In" and "Out" of data, as already anticipated above: when data are invoked by favoring their emergence as cognitive chances and when they are further manipulated, it is clear that abductive cognition tends to produce results that go beyond data, in which creativity of deep hypotheses dominates.

In this perspective the proper experimental test involved in the Peircean evaluation phase, which for many researchers reflects in the most acceptable way the idea of abduction as inference to the best explanation, just constitutes a *special* subclass of the process of the adoption of the abductive hypothesis—the one which involves a terminal kind (*iv*) of actions (experimental tests), and should be considered ancillary to the nature of abductive cognition, and inductive in its essence. We have indeed to remark again that in Peirce's mature perspective on abduction as embedded in a cycle of reasoning, induction just plays an evaluative role.

3.2.1 An Example: Construals Create Epistemological Value "In" and "Out" of Data

At the end of this section is mandatory to present some examples of manipulative abduction, that obviously involve model-based aspects. In this kind of *action-based* abduction the suggested hypotheses are inherently ambiguous until articulated into configurations of real or imagined entities (data, images, models or concrete apparatus and instruments). In these cases only by experimenting, can we discriminate between possibilities: they are articulated behaviorally and concretely by manipulations and then, increasingly, by words and model-based aspects such as pictures and various kinds of simulations. Gooding [23] refers to this kind of concrete manipulative reasoning when he illustrates the role in science of the so-called "construals" that embody tacit inferences in procedures that are often apparatus and machine based. In these cases we face a process of attribution of epistemic values to "data", which become the product of cognitive embodiments together with an expert manipulation of objects in a highly constrained experimental environment, embodiments that are directed by abductive movements that imply the strategic application of old and new *templates* of behavior also connected with extra-theoretical components, for instance emotional, aesthetical, ethical, and economic.[8]

---

[8] Tweney [61] has emphasized the importance of externalized cognitive artifacts used in the service of the "seeing data" of scientists. In turn they are distributed "in the strong sense that not all of

The hypothetical character of construals is clear: they can be developed to examine further chances, or discarded, they are provisional creative organization of experience and data and some of them become in their turn hypothetical *interpretations* of experience and data that are in this way further redesigned and adjusted, that is more theory-oriented, their reference is gradually stabilized in terms of established observational practices. Step by step the new interpretation—that at the beginning is completely "practice-laden"—relates to more "theoretical" modes of understanding (narrative, visual, diagrammatic, symbolic, conceptual, simulative), closer to the constructive effects of abduction. When the reference is stabilized the effects of incommensurability with other stabilized observations can become evident. But it is just the construal of certain phenomena as data that can be shared by the sustainers of rival theories. Gooding [23] shows how Davy and Faraday could see the same attractive and repulsive actions at work in the phenomena they respectively produced; their discourse and practice as to the role of their construals of phenomena clearly demonstrate they did not inhabit different, incommensurable worlds in some cases. Moreover, the experience is constructed, reconstructed, and distributed across a social network[9] of negotiations among the different scientists by means of construals.[10]

Gooding introduces the so called *experimental maps*[11] that are the epistemological two-dimensional tools that we can adopt to illustrate the conjecturing (abductive) role of actions from which scientists "talk and think" about the world. They are particularly useful to stress the attention to the interaction of hand, eye, and mind inside the actual four-dimensional scientific cognitive process. The various procedures for manipulating data, objects, instruments and experiences will be in their turn reinterpreted in terms of procedures for manipulating concepts, models, propositions, and formalisms. Scientists' activity in a material environment first of all enables a rich perceptual experience that has to be reported mainly as a visual experience by means of the constructive and hypothesizing role of the experimental narratives.

---

the agentive movement of thought is localized solely within an individual skin". I think a further light on the role of construals is shed by Franklin who usefully analyzes the so-called "exploratory experiments" that prior to theorizing investigate the world "without premature reflection of any great subtlety", like Bacon says [4, p. 210], and where there is no particular hypothesis being pursued. They serve "[…] to find interesting patterns of activity from which the scientists could later generate a hypothesis" [18, p. 894]

[9] Cf. [55, 60].

[10] Gooding [24] further analyze the role of various kinds of experiments on data—ranging from the idealized crucial ones to those that are exploratory and/or controversial—like mediating models in the framework of an agent-based approach. Every agent or actor can investigate a world of experiments and other agents, in a setting where eventually scientists invent and negotiate ways of representing aspects of the world they are investigating. The process is "adaptive" and "inherently social" and inference is seen as a continuous activity of belief-revision where the distributed and collaborative aspects are acknowledged. In this perspective experiments are seen as mediating between at least four sets of objects: hypotheses, procedures, physical setups and observable outcomes.

[11] Circles denote concepts (mentally represented) that can be communicated, squares denote things in the material world (bits of apparatus, observable phenomena, data) that can be manipulated—lines denote actions.

The construals aim at arriving to a shared understanding overcoming all conceptual conflicts. As I said above they constitute a provisional creative organization of experience: when they become in their turn hypothetical interpretations of experience, that is more theory-oriented, their reference is gradually stabilized in terms of established and shared observational practices that also exhibit a cumulative character. It is in this way that scientists are able to communicate the new and unexpected information acquired by experiment and action.

To illustrate this process—from manipulations, to narratives, to possible theoretical models (visual, diagrammatic, symbolic, mathematical)—we need to consider some observational techniques and representations made by Faraday, Davy, and Biot concerning Oersted's experiment about electromagnetism. They were able to create consensus because of their conjectural representations that enabled them to resolve phenomena into stable perceptual experiences and respective data. Some of these narratives are very interesting. For example, Faraday observes: "[...] it is easy to see how any individual part of the wire may be made attractive or repulsive of either pole of the magnetic needle by mere change of position [...]. I have been more earnest in my endeavors to explain this simple but important point of position, because I have met with a great number of persons who have found it difficult to comprehend". Davy comments: "It was perfectly evident from these experiments, that as many polar arrangements may be formed as chord can be drawn in circles surroundings the wire". Expressions like "easy to see" or "it was perfectly evident" are textual indicators inside the experimental narratives of the stability of the forthcoming interpretations. Biot, in his turn, provides a three-dimensional representation of the effect by giving a verbal account that enables us to visualize the setup: "suppose that a conjunctive wire is extended horizontally from north to south, in the very direction of the magnetic direction in which the needle reposed, and let the north extremity be attached to the copper pole of the trough, the other being fixed to the zinc pole [...]" and then describes what will happen by illustrating a sequence of step in a geometrical way:

> Imagine also that the person who makes the experiment looks northward, and consequently towards the copper or negative pole. In this position of things, when the wire is paced above the needles, the north pole of the magnet moves towards the west; when the wire is placed underneath, the north pole moves towards the east; and if we carry the wire to the right or the left, the needle has no longer any lateral deviation, but is loses its horizontality. If the wire be placed to the right hand, the north pole rises; to the left, its north pole dips [...].[12]

It is clear that the possibility of "seeing" interesting things through the experiment depends from the manipulative ability to get the correct information and to create the possibility of a new interpretation (for example a simple mathematical form) of electromagnetic natural phenomena, so joining the theoretical side of abduction. Step by step, we proceed until Faraday's account in terms of magnetic lines and curves.

---

[12] The quotations are from [16, p. 199], [12, pp. 282–283] and [5, pp. 282–283], cited by [23, pp. 35–37].

3.3   Asking "Why" Means to Reach Beyond Data: Irrelevance and Implausibility Trigger Creativity

By adopting the received logical view of abduction related to the traditional problem of diagnostic reasoning, expressed by the so-called AKM-schema[13] we restrict our general idea of abduction being conditioned by special cases: unfortunately, by restricting our perspective in this way, some good solutions of an abductive problem that would be remarkable and productive in other cases (for example different with respect medical diagnosis, such as in the case of scientific discovery, in which asking "why" means to reach beyond data in a deeper way) can be ruled out. To avoid this result the GW-schema and the EC-Model of abduction I have described in the previous sections can be of some help. The GW-schema does not refer to consistency and minimality as necessary requirements: in the first case we have to remember that knowledge bases not only are incomplete but often incorporate inconsistencies; in the second one there is not compelling need to consider that a solution for an abductive problem should be one with an assigned length.

However, it has to be said that many standard perspectives on abduction still demand two properties, which are presented as possessed by "every" kind of solution for an abductive problem:

1. *Relevance*: the solution, the guessed hypothesis H, should be relevant to the problem: for example, if an agent's knowledge does not suffice to know why the bartender in Kuala Lumpur has been killed, releasing the true Newtonian hypothesis that the planets move according to the law of gravitation has nothing to do with the given problem: it is not relevant.
2. *Plausibility*: The abduced hypothesis H should be characterized by some designated degree of plausibility. If an agent's knowledge does not suffice to know who killed the bartender in Kuala Lumpur, releasing the hypothesis that the killer is the President of United States has to do with the problem (because, after all, the President is a human being and we know human beings are potential killers) but is sufficiently implausible as to count as a solution.

Indeed in the eco-cognitive perspective the relevance of a guessed hypothesis would seem a trivial requirement, because it is "hard to see how it might fail to be relevant", as Estrada-González says [14, p. 185]. He further adds: "Someone might press the point that what is required is the relevance not of solutions, but of candidates to be solutions. However, I think it might go against all those pleas connecting abduction with creativity, hypothesis generation, guessing, etc." I basically agree with him. This is exactly the point to be stressed and further explained. First of all we have to note that relevance is *context-* and *time-dependent*. As already said above

---

[13] The AKM-schema involves for the *consistency* and *minimality* constraints. Indeed The classical schematic representation of abduction is expressed by what [20] call AKM-schema, which is contrasted to their own (GW-schema), I described above. For *A* they refer to Aliseda [1, 2], for *K* to Kowalski [33], Kuipers [34], and Kakas et al. [32], for *M* to Magnani [40] and Meheus [54]. A detailed illustration of the AKM schema is given in [41, Chapter two, Sect. 2.1.3].

in Sect. 3.1.2, when Feyerabend [17] emphasizes the role of what he calls "counterinduction", he is just presenting to us the complete unreasonable and unwarranted character of scientific discovery: the guessed hypothesis could be devoid of relevance to the problem in the framework of the upholders of the rival theory but also, even if not necessarily, in the perspective of the agent herself that—paradoxically—guessed the new "strange" hypothesis. Of course the relevance requirement is related to the current state of knowledge of both agonists. However, the new hypothesis can result "relevant" later on, for example when recognized as a genuine new discovery. To summarize, candidates to be solutions which seem weird—irrelevant—soon can become relevant if they are recognized as solutions.

Something similar can be said in the case of *plausibility*. First of all, in general we do not necessarily have to be obsessed by the plausibility of our guessed hypotheses (even if we know that looking for plausibility is a human good and wise heuristic), an implausible hypothesis can later on result plausible. Moreover, when a hypothesis solves the problem at hand, this is enough as to count as solution of the abductive problem (even if, not necessarily a *good* solution or the *best* solution). If we want to preserve the property of plausibility, at most we can say that in some cases it is just *potential*, given the time-dependency I have illustrated.

To make an example, the strange Cartesian hypothesis of a plenum vortices made of particles, destroyed by the Newtonian concept of action at distance, later on appeared more rational and fully compatible with the Einsteinian framework:

> Thus Descartes was not so far from the truth when he believed he must exclude the existence of an empty space. The notion indeed appears absurd, as long as physical reality is seen exclusively in ponderable bodies. It requires the idea of the field as the representative of reality, in combination with the general principle of relativity, to show the true kernel of Descartes' idea; there exists no space "empty of field" [13, pp. 375–376].

In sum, irrelevance and implausibility not always are offensive to reason and incline reasoning, very far beyond the rough data, to deep and new creative results[14]: to delineate the fill-up problem neither relevance not plausibility are necessary, they are just two "typical" smart and fruitful principles human beings subjectively adopt to look for hypotheses. Unfortunately, they are no longer "typical", for example, in the case of high-level kinds of cognitive creativity. Also the GW-schema acknowledges the fact that relevance or plausibility cannot be taken to be general conditions on hypothesis-selection.

---

[14] Scientists would agree that the really surprising and fruitful thought arising from abduction has to challenge prevailing conceptions by suggesting ideas that are *prima facie* neither relevant nor plausible, or that even appear to contradict pre-established notions (see e.g. the book by Livio [35]).

3.4 Beyond Learning from Data: Trans-Paradigmatic and Transepistemic Abductions

When I said in Sect. 3.1.2 that abduction can be knowledge-enhancing I was referring to various types of new produced knowledge of various novelty level, in absence of an empirical evaluation phase, or inductive phase, as Peirce called it. Some cases of new knowledge produced in science (for example, conventions and intermediate models used in research settings), are cases of knowledge enhancing abduction. However, also knowledge produced in an artificial game thanks to a smart application of strategies or to the invention of new strategies and/or heuristics has to be seen as the fruit of knowledge enhancing abduction. I have noted that this means that abduction is not necessarily ignorance-preserving (reached hypotheses would always be "presumptive" and to be accepted they always need empirical confirmation), as contended by Gabbay and Woods (see [62]).[15] Abduction can creatively build new knowledge by itself (that is an inference not necessarily characterized by an empirical evaluation phase, or inductive phase), as various examples coming from the area of history of science and other fields of human cognition clearly show.

I contend that to reach selective or creative good abductive results,"beyond data", efficient strategies have to be exploited, but it is also necessary to count on an environment characterized by what I have called *optimization of eco-cognitive situatedness*, in which that eco-cognitive openness already envisaged by Aristotle thanks to the emphasis on "leading away"[16] is fundamental. To favor good creative and selective abduction reasoning strategies must not be "locked" in an external restricted eco-cognitive environment such as in a scenario characterized by fixed definitory rules and finite material aspects (an artificial game, Go or Chess, for example), which would function as cognitive mediators able to constrain agents' reasoning.[17] In brief, the optimization of eco-cognitive situatedness concerns the substantial problem of *discoverability* and *diagnosticability*, almost totally disregarded in the literature on abduction (and just sketched by Peirce himself).[18]

Research on abduction has already implicitly emphasized the fruitful role of cognitive openness. Hendricks and Faye [25] consider trans-paradigmatic abduction a form of discovery in which a guessed hypothesis transcends the prompt empirical agreement between two paradigms. The paradigms are presumed to belong to the same field (for example physics) where one of the fields is well established and the other is emerging (for example classical and quantum physics):

---

[15] Woods has recently enriched, modified, and moderated his views of ignorance-preservation, see [63].

[16] Cf. section above Sect. 3.2.

[17] More details concerning the role of locked and unlocked strategies are illustrated in the recent [49] and below in the last Sect. 3.7.

[18] I plan to devote part of my future research to study these aspects of abductive cognition. My recent book *Discoverability. The Urgent Need of an Ecology of Human Creativity*, Springer, Cham, Switzerland, 2022, analyzes these aspects of abductive cognition.

A case in point would be the formulation of the hypothesis of electron spin. Bohr considered the spin conjecture as a welcome supplement to the current magnetic core theory. Pauli remained rather skeptical pertaining to the spin hypothesis due to the fact that it actually required the theory of quantum mechanics for its proper justification, which was not part of the background knowledge at the time of the conjecture. In such cases two paradigms are competing and the abduction is then dependent upon whether the conjecture is made within the paradigm or outside it. Hence we distinguish between paradigmatic and transparadigmatic abduction (cit., p. 287).

Furthermore, people draw on different domains of knowledge to arrive to an abductive conclusion thanks to what [22] call "transepistemic abduction" (TeA), which illustrates how two agents, in order to successfully explain a phenomenon, reason across two very distant cognitive fields (for example computational and psychosocial domains) despite each agent being ignorant of the other domain knowledge. The authors themselves acknowledge that TeA represents a case that is concerned with my eco-cognitive perspective: "TeA may not necessarily accommodate wider understandings of abduction like the eco-cognitive model proposed by Magnani. For example, TeA may not necessarily encompass perceptions, aesthetic decisions or moral judgements in the way that a eco-cognitive view of abduction might" [22, p. 473].

Another interesting procedure that can refer to higher abductive processes in need of cognitive openness is the chunk-and-permeate method [6], in which consideration is given to conditions under which mutually incompatible well-grounded theories can interact to bring forth solutions to problems which neither theory can solve on its own. This method introduces a paraconsistent reasoning strategy, in which information is broken up into chunks, and a limited amount of information is allowed to flow between chunks, and it is applied to model the reasoning employed in the original infinitesimal calculus.

It is now useful to provide a short introduction to the concept of eco-cognitive openness from a logical point of view. The new perspective inaugurated by the so-called *naturalization of logic*[19] contends that the normative authority claimed by formal models of ideal reasoners to regulate human practice on the ground is, to date, unfounded. It is necessary to propose a "naturalization" of the logic of human inference. Woods holds a naturalized logic to an adequacy condition of "empirical sensitivity" [62]. A naturalized logic is open to study many ways of reasoning that are typical of actual human knowers, such as for example fallacies, which, even if not truth preserving inferences, nonetheless can provide truths and productive results. Of course one of the best examples is the logic of abduction, where the naturalization of the well-known fallacy "affirming the consequent" is at play. Gabbay et al. [20, p. 81] clearly maintain that Peirce's abduction, depicted as both (a) a surrender to an idea, and (b) a method for testing its consequences, perfectly resembles central aspects of practical reasoning but also of creative scientific reasoning.

It is in the spirit of this project of naturalization of logic that my recent research on abduction [45, 47] stressed the importance, beyond data, of various types of that "good" abductive cognition as they are occurring thanks to the already quoted

---

[19] I have illustrated this new project in [44].

optimization of situatedness, that is abductive cognition in a situation of strong eco-cognitive openness. This is for example very important in scientific reasoning because it refers to that activity of creative hypothesis generation which characterizes one of the more valued aspects of rational knowledge. The study above teaches us that situatedness is related to the so-called eco-cognitive aspects, referred to various contexts in which knowledge is "traveling": to favor the solution of an inferential problem—especially in science but also in the case of other abductive problems, such as diagnosis—the richness of the flux of information has to be maximized.

3.5   Can Models Computationally Derived from Data Provide Scientific Knowledge of Physical and Biological Systems?

It is important to note that, in the traditional view offered by Turing, the digital machine (a discrete state machine) is certainly an alphabetic machine: its conditions of possibility resort to human evolution towards alphabetic natural language. Longo [36] contends that this fact is at the origin of that tremendous "discretization of knowledge" that the Turing's achievements have created. The "continuous" natural language is indeed transmuted by the alphabet in something separated into small atoms, which forge letters. These atoms do not present any kind of meaning that instead comes out thanks to their syntactical aggregation made by skilled human agents able to sensibly combine them. This discreteness, typical of digital machines, is the fundamental aspect that motivates their *imitation* power—they are mimetic machines, *mimetic minds*, as I say[20]—and Turing himself contrasted this simple imitation power to the much stronger epistemological power of the *modeling* capacity of mathematics, when he was thinking about the science of morphogenesis.

At this point, the problem of imitation leads us to consider a further aspect of physical computation that affects the problem of abductive cognition from data, in which the epistemological consequences of the more of less distance from them is at stake: when computers are further used to model physical (or biological systems), for example massively starting from big data, are we still dealing with imitation or with a kind of reliable production of knowledge that could occasionally be called "scientific"? Let us see more details concerning this problem.

As I have illustrated in [48] physical computation is a transposition of physical evolution for abstract computation, that is, we can say, following Turing, we have "educated"(or "domesticated", as I say)[21] a physical system to perform a computation (and consequently a related cognitive process). Once this task is performed, we can "in turn" submit a physical (or biological) dynamic of a system to a computational modeling, that is, we can use computers to simulate the behavior of a physical or

---

[20] Cf. [41, Chapter three].

[21] Cf. [51]. I have recently systematized these issues in my book *Eco-Cognitive Computationalism. Cognitive Domestication of Ignorant Entities*, Springer, Cham, Switzerland, 2022.

a biological system. In the meantime, it is important to note that in this case the computational simulator and the physical or biological system simulated interact at the abstract level and what is simulated is a *model* of the physical or biological system, not the system itself ([28], p. 17). We know that computational modeling of physical or biological processes can be extremely useful as a heuristic tool in actual scientific research, also at the creative level (just to make a simple example, to simulate the behavior—through modeling[22]—of a physical system during an experiment), but we have to note that the study of a physical or a biological evolution cannot always take advantage of a computational representation. Hence, what is the epistemological status of computational simulation?

The dichotomy between discreteness and continuity involves a reflection upon the other related dichotomy between imitation (as an effect of the computational representation) and intelligibility (as the fruit of scientific knowledge). I will devote the next paragraphs to better describe this important issue, at least from a general theoretical point of view. I have said above that the digital machine (a discrete state machine) is first of all an alphabetic machine, made possible thanks to the human evolution to alphabetic natural language (of course, we know it is also based on the so-called logical and formal machine). As I have already noted, this fact explains that powerful "discretization of knowledge" that mainly characterizes the "computational turn".

Notwithstanding the triumph of discretization, in western written natural languages but also in philosophical and logical knowledge, from Democritus to Descartes and from the modern XIX century axiomatics to the computational turn, we are still facing the conundrum represented by the fact that, however, this simple reality of small components is actually very *complex*, as recent scientific research into the dynamical systems theory, quantum physics, and biology demonstrates. For example, it is difficult to study the cell only by referring to its constituents, and also its "wholeness" is fundamental. The suspect is that, when we use computational devices from data, which are discrete machines, to produce knowledge about physical and biological systems, some serious expressive limits arise: it is unlikely that these machines can play the role of instruments for directly building scientific intelligibility regarding complex objects/systems such as physical and biological entities (and also human cognitive systems themselves).

The deep difference between the idea of scientific knowledge as simple *imitation*, as developed by digital machines and their "modeling capacity", and the idea of scientific knowledge as generation of *rational intelligibility* is a core problem of the dynamical approach [58]. As I already said above, Turing himself emphasized the difference between the simple *imitation* capacity of machines and the *modeling* power of mathematics: the double pendulum, which is a perfect deterministic machine, only expressed by two equations, is sensible to minor variations, below the threshold of observability. It is a typical chaotic deterministic system, and it is extremely difficult to represent it by a mimetic machine [36]. Longo also notes that this is a system sensitive to initial conditions. It can instead described using the mathematics of

---

[22] Cf. [53].

deterministic chaos, in which determination does not involve predictability: in this system, a process does not follow the same trajectory even if we reiterate it with the same initial conditions, within the limits of physical measures (*ibid.*). Furthermore, we can also usefully observe that, in the case of Turing machines that simulate such systems, when we restart after having processed analog simulations, using the same initial data, the same already seen trajectory is performed and, on the contrary, the "real" pendulum behavior is completely different, when we restart the pendulum never performs the same trajectory.

Only "mathematical" models can explain the structure of physical causality of the a system at play, and digital simulation from data only *resembles* or *imitates* causality, always realizing Laplacian and predictable processes. The explanation of this conundrum resorts to the fact that, at the roots of digital data, there is a *discrete* topology, but physical measurement is always an interval that is optimally represented by continuous mathematics. Even if sometimes the use of mathematical equations in physics does not have a predictive power, this provides a knowledge characterized by a rich *qualitative* epistemological value, able to make intelligible the physical causality of the system [37]. Finally, we have to note that, with respect to physics, in the case of biological organisms, the gap between simulation and intelligibility is even worse because the variability is dominant.[23]

In summary, digital simulation from data—even if epistemological useful at the level of intermediate modeling during the processes of scientific research and discovery, as I have already indicated above in this section—produces an epistemologically distorting result due to its simply mimetic quality.

3.6   Can Invasive Computational Learning from Big Data Jeopardize Human Creative Abduction in Science?

Some computational research based on the exploitation of big data and of deep learning systems, which is surely also directed by groups that aim at promoting a related business, also seems to challenge what we can call "epistemic integrity", for example trivially contending the capacity to replace with artifacts human scientific creative abduction. To make a paradigmatic example Calude and Longo contend that

> Very large databases are a major opportunity for science and data analytics is a remarkable new field of investigation in computer science. The effectiveness of these tools is used to support a "philosophy" against the scientific method as developed throughout history. According to this view, computer-discovered correlations should replace scientific understanding as a guide to prediction and action. Consequently, there will be no need to give scientific meaning to phenomena, by proposing, say, causal relations, since regularities in very large databases

---

[23] In [38, p. 1] Longo contends that, "in biology, in particular, the introduction of information as a new observable on discrete data types has been promoting a dramatic reorganization of the tools for knowledge" and that some consequences of this effect have been induced in life sciences, with particular emphasis on research on cancer.

are enough: "with enough data, the numbers speak for themselves". The "end of science" is proclaimed [7, p. 595].

Calude and Longo demonstrate, taking advantage of deep classical results from ergodic theory, Ramsey theory, and algorithmic information theory, how absurd is this contention and that instead very large database[24] present too many arbitrary—and so spurious—correlations, which surely cannot be considered examples of pregnant scientific creative abduction, but just uninteresting generalizations,[25] even if made thanks to sophisticated artifacts: "Too much information tends to behave like very little information" (*ibid.*) This study is the correct answer to the implicit challenge to scientific cognition proposed in June 2008 by C. Anderson, former editor-in-chief of *Wired Magazine*, who wrote an article titled "The end of theory: the data deluge makes the scientific method obsolete" contending that "with enough data, the numbers speak for themselves", science as we know will be replaced by robust correlations in immense databases![26]

Amazing examples of the use of these correlations are illustrated. The following is an eloquent example that indicates the extreme spontaneity and naturalness of some types of epistemic irresponsibility, especially in the case of the cognitive performances of rudimentary politicians and mass media journalists:

A 2010 study conducted by Harvard economists Carmen Reinhart and Kenneth Rogoff reported a correlation between a country's economic growth and its debt-to-GDP ratio. In countries where public debt is over 90 percent of GDP, economic growth is slower, they found. The study gathered a lot of attention. Google Scholar listed 1218 citations at the time of writing, and the statistic was used in U.S. economic policy debates in 2011 and 2012 [...]. The study isn't conclusive, though-in fact, it's far from it. As noted by John Irons and Josh Bivens of the Economic Policy Institute, it's possible that the effect runs the other way round, with slow growth leading to high debt. Even more worryingly, the research didn't hold up on replication. But by the time that became clear, the original study had already attracted widespread attention and affected policy decisions [7].

The authors also note that European policy makers largely referred to that paper till 2013. The EU Commissioner for Economic Affairs—2009–2013—referred to

---

[24] A special issue on the scientific method and the machine learning approach to big data analysis has been published by the *International Journal of Knowledge Society Research* [10]. Moreover, [59] provides a discussion of the so-called libertarian-inspired "finders, keepers" ethics, that is the curious "morality" adopted by the business model of big data companies, which can "legitimately" appropriate (the fruits of) their results. Frizzo-Barker [19] also describe large-scale networked genetic material as a disruptive technology. On one hand, clinical genomics advances life-saving innovation through precision medicine. On the other, the digital databases they are built upon raise new concerns for informational risk to personal privacy.

[25] Such as the relation between the orientation of a comet's tail and the Emperor's chances of a military victory. Large database seem to rejoin magic, where underlying rational causes are lacking: but correlations do not supersede causation.

[26] More and more often researchers in different disciplines have been interested in using big data for their applications. Nonetheless, "For big data, spurious correlation refers to uncorrelated variables being falsely found to be correlated due to the massive size of the dataset" [21, p. 143]. Studies that use big data are thus more likely to achieve the "generalizability" criterion of the findings, yet they are also prone to potentally produce uninteresting and/or insignificant, though generalizable, results due to the mentioned spurious correlations (see e.g. [15]).

the Reinhart–Rogoff correlation as a key guideline for his past and present economic views and the British Chancellor of the Exchequer—since 2010—observed in April 2013 that Rogoff and Reinhart demonstrate "convincingly" that all financial crises have their ultimate origins in the public debt.

A strong criticism of another invasive computational "subculture" provided by [39], addresses the value of the still widespread metaphor "DNA is a program" (the programming paradigm), used both in molecular biology and in its popularization. The authors contend that the metaphor and the model are fundamentally inadequate not only in biology but also from the point of view of both physics and computer science. Still in this case, analogously to the case of big data I have just illustrated, the programming paradigm is not theoretically sound as a causal framework for relating the genome to the phenotype, a much more complex process than the one which is depicted by an explanation based on this computational paradigm.

3.7 The Triumph of Shallow Massive Data-Driven Knowledge: Deep Learning and Locked Strategies

Let me describe another challenge to human abductive reasoning performed in the area of artificial intelligence (AI), very interesting but exaggeratedly emphasized by mass-media, social networks, and the corporation itself, which proudly made it, that is Google. A long tradition in artificial intelligence—mainly in the area usually called "machine learning"—concerned the epistemologically very impressive computational AI applications that involve the abductive processes in scientific discovery and mathematical reasoning and   creativity.[27]

The example I am describing here regards an AI program that was able to play the Game Go very successfully, and so not involved in simulating scientific discovery but certainly human skillful strategic ability and creativity. It is well-known that in 2015 Google DeepMind's program AlphaGo beat Fan Hui, the European Go champion and a 2 dan (out of 9 dan possible) professional, five times out of five with no handicap on a full size 19x19 board. In March 2016, Google also challenged Lee Sedol, a 9 dan considered the top world player, to a five-game match. The program shot down Lee in four of the five games. It seems the looser acknowledged the fact the program adopted one unconventional move—never played by humans—leading to a new strategy, so performing a very "human" capacity, and I have to say, better than the one of the more skilled humans. AlphaGo learned to play the game by checking data of thousands of games, and may be also those played by Lee Sedol, exploiting the so-called "reinforcement learning", which means the machine plays against itself to further enrich and adjusts its own neural networks based on trial and error. Of course the program also implicitly performs what we call "reasoning

---

[27] I have discussed the problem of automatic scientific discovery with AI programs in [41, Chapter two, Sect. 2.7 "Automatic Abductive Scientists"].

strategies" to reduce the search space for the next best move from something almost infinite to a more calculable quantity.

Cohleo and Thompsen Primo, de facto testify in a passage I will report soon that for an AI program as AlphaGo is relatively easy to reproduce at the computational level what I have called *locked reasoning strategies* [46].[28] These programs are characterized by locked abductive strategies: they deal with weak (even if sometimes amazing) kinds of hypothetical creative reasoning, because they are limited in what I called in this chapter (see above Sect. 3.4) eco-cognitive openness, which instead qualifies human cognizers who are performing higher kinds of abductive creative reasoning, where cognitive strategies are instead *unlocked*. What counts here is that in the human creative inferential processes the strategies are unlocked because, even if local constraints are always at play in the interaction humans/environments, no predetermined background is established. On the contrary, what happens in the case of human made "artificial games" such as Go, or in the case of their computational counterpart, such as AlphaGo/AlphaZero?[29] In these two last cases, the involved cognitive strategies are *locked*, as I will describe in the following paragraphs. In summary, a kind of general reason of this simplicity would be that this kind of human reasoning is less creative than others, even if it is so spectacular and performed in an optimal way only by very skilled and intelligent subjects.

Here the interesting comments by Cohleo and Thompsen:

> Let us compare the key ideas behind Deep Blue (Chess) and AlphaGo (Go). The first program used values to assess potential moves, a function that incorporated lots of detailed chess knowledge to evaluate any given board position and immense computing power (brute force) to calculate lots of possible positions, selecting the move that would drive the best possible final possible position. Such ideas were not suitable for Go. A good program may capture elements of human intuition to evaluate board positions with good shape, an idea able to attain far-reaching consequences. After essays with Monte Carlo tree search algorithms, the bright idea was to find patterns in a high quantity of games (150,000) with deep learning based upon neural networks. The program kept making adjustments to the parameters in the model, trying to find a way to do tiny improvements in its play. And, this shift was a way out to create a policy network through billions of settings, i.e., a valuation system that captures intuition about the value of different board position. Such search-and-optimization idea was cleverer about how search is done, but the replication of intuitive pattern recognition was a big deal. The program learned to recognize good patterns of play leading to higher scores, and when that happened it reinforces the creative behavior (it acquired an ability to recognize images with similar style) [9, p. 21].

We humans with our organic brains do not have to feel humiliated by these bad news...Human portentous performances with the game Go and other human ways of reasoning, even more creative than the ones involved in a locked strategic reasoning, cannot reach the global echo AlphaGo/AlphaZero gained. The reason is simple,

---

[28] In this case the strategies which are activated are multiple but all are "locked" because, as I will better explain below in Sect. 3.7.1, the components of each scenario in which they are applied are always the same (for example in the case of the game Go just the number of present stones and their configurations change), in a finite and unchanging framework (no new rules, no new objects, no new boards, etc.).

[29] AlphaGo Zero is a version of DeepMind's Go software AlphaGo; the recent AlphaZero further enriches AlphaGo Zero and learns by its own played games.

human-more-skillful-abductive creative performances—still cognitively gorgeous—
are not sponsored by Google, which is a powerful corporation that can easily obtain
a huge attention by aggressive media, a lot of internet web sites, and social networks enthusiast ignorant followers, more easily impressionable by the "miracles"
of AI, robotics, and in general, information technologies, than by exceptional human
knowledge achievements, always out of their material and intellectual reach.

Google managers also believe AI programs similar to AlphaGo/AlphaZero could
be used to help scientists solve tough real-world problems in healthcare and other
areas. This is more than welcome. Of course I guess Google will also expect to
implement some business thanks to a commercialization of new AI capacities to
gather information and making abductions on it. Marketing aims are always important
in these cases.

The Wikipedia entry DeepMind (https://en.wikipedia.org/wiki/DeepMind, date
of access 23 July 2022) [DeepMind is a British AI company founded in September
2010 and acquired by Google in 2014, the company realized the AlphaGo program]
reports the following non contested passage:

> In April 2016 New Scientist obtained a copy of a data-sharing agreement between Deep-
> Mind and the Royal Free London NHS Foundation Trust, which operates the three London
> hospitals which an estimated 1.6 million patients are treated annually. The agreement shows
> DeepMind Health had access to admissions, discharge and transfer data, accident and emer-
> gency, pathology and radiology, and critical care at these hospitals. This included personal
> details such as whether patients had been diagnosed with HIV, suffered from depression
> or had ever undergone an abortion in order to conduct research to seek better outcomes in
> various health conditions. A complaint was filed to the Information Commissioner's Office
> (ICO), arguing that the data should be pseudonymised and encrypted. In May 2016, New Sci-
> entist published a further article claiming that the project had failed to secure approval from
> the Confidentiality Advisory Group of the Medicines and Healthcare products Regulatory
> Agency. In May 2017, Sky News published a leaked letter from the National Data Guardian,
> Dame Fiona Caldicott, revealing that in her "considered opinion" the data-sharing agreement
> between DeepMind and the Royal Free took place on an "inappropriate legal basis". The
> Information Commissioner's Office ruled in July 2017 that the Royal Free hospital failed
> to comply with the Data Protection Act when it handed over personal data of 1.6 million
> patients to DeepMind.

Academic epistemologists and logicians have to monitor the exploitation of
these AI and of other computational tools, which could present uses that can be
less transparent than the simple and clear—and so astonishing—performance of
AlphaGo/AlphaZero in games against humans. Good software, which represents a
great opportunity for science and data analytics, can be transformed in a tool that does
not respect epistemological rigor and/or basic western received moral standards.[30]

---

[30] On the spurious correlations generated by the management of big data cf. above Sect. 3.6.

3.7.1 Reading Ahead

A fundamental strategy we immediately detect in artificial games such as Go, which is necessary for proficient and smart tactical play, is the capacity to *read ahead*, as the Go players usually say. Reading ahead is a practice of generating groups of anticipations that aim at being very robust and complex (either serious-minded or intuitive) and that demand the consideration of

1. Clusters of moves to be adopted and their potential outcomes. The available scenario at time t_1, exhibited by the board, represents an adumbration[31] of a subsequent potential more profitable scenario at time t_2, which indeed is abductively credibly hypothesized: in turn, one more abduction is selected and actuated, which—consistently and believably—activates a particular move that can lead to an envisaged more fruitful scenario.
2. Possible countermoves to each move.
3. Further chances after each of those countermoves. It seems that some of the smarter players of the game can read up to 40 moves ahead even in hugely complex positions.

In a book published in Japan, related to the description of various strategies that can be exploited in Go games, Davies emphasizes the role of "reading ahead":

> The problems in this book are almost all reading problems. [...] they are going to ask you to work out sequences of moves that capture, cut, link up, make good shape, or accomplish some other clear tactical objective. A good player tries to read out such tactical problems in his head before he puts the stones on the board. He looks before he leaps. Frequently he does not leap at all; many of the sequences his reading uncovers are stored away for future reference, and in the end never carried out. This is especially true in a professional game, where the two hundred or so moves played are only the visible part of an iceberg of implied threats and possibilities, most of which stays submerged. You may try to approach the game at that level, or you may, like most of us, think your way from one move to the next as you play along, but in either case it is your reading ability more than anything else that determines your rank ([11] p. 6).

Further strategies that are usefully adopted by human players in the game Go are for instance related to "global influence, interaction between distant stones, keeping the whole board in mind during local fights, and other issues that involve the overall game. It is therefore possible to allow a tactical loss when it confers a strategic advantage".[32]

The material and external scenarios which are composed by the sensible objects—stones and board—that characterize artificial games are the fruit of a cognition "sedimented"[33] in their embodiment, after the starting point of their creation and subsequent uses and modifications. The cognitive tools that are related to the application of

---

[31] The word belongs to the Husserlian philosophical lexicon [29] I have analyzed in its relationship with abduction in ([41] Chap. 4).

[32] Cf. Wikipedia, entry Go (game) https://en.wikipedia.org/wiki/Go_(game) date of access 23 July 2022.

[33] An expressive adjective still used by Husserl [30], translated by D. Carr and originally published in *The Crisis of European Sciences and Transcendental Phenomenology* [1954].

both the game allowed rules and the individual inferential talents owned by the two players, strategies, tactics, heuristics, etc. are sedimented in those material objects (artifacts, in this case) that become *cognitive mediators*:[34] for example they orient players' inferences, transfer information, and provoke reasoning chances. Once represented internally, the external subsequent scenarios become object of mental manipulation and new ones are further made, to the aim of producing the next most successful move.

It is important to note again that these strategies, when actuated, are certainty characterized by an extended variety, but all are "locked", *because the elements of each scenario are always the same* (what changes is merely the number of seeable stones and their dispositions in the board), in a finite and stable framework (no new rules, no new objects, no new boards, etc.). These strategies are devoid of the following feature: they are not able to recur to reservoirs of information *different* from the ones available in the fixed given scenario. It is important to add a central remark: of course the "human" player can enrich and fecundate his strategies by referring to internal resources not necessarily directly related to the previous experience with Go, but with other preexistent skills belonging to disparate areas of cognition. This is the reason why we can say that the strategies of a "human" player present a less degree of closure with respect to the automatic player AlphaGo/AlphaZero. In humans, strategies are locked with respect to the external rigid scenario, but more open with respect to the mental field of reference to previous wide strategic experiences; in AlphaGo/AlphaZero and in deep learning systems, the strategic reservoir cannot—at least currently—take advantage of that mental openness and flexibility typical of human beings: the repertoire is merely formed/learned to play the game by checking data of thousands of games, and no other sources.

I have also to say that the notion of cognitive locked strategy I am referring to here is not present in and it is unrelated to the usual technical categorizations of game theory. Fundamentally, in combinatorial game theory, Go can be technically illustrated as zero-sum (player choices do not increment resources available-colloquially), perfect-information, partisan, deterministic strategy game, belonging to the same class as chess, checkers (draughts) and Reversi (Othello). Moreover, Go is bounded (every game has to end with a victor within a finite number of moves and time), strategies are obviously associative (that is in function of board position), format is of course

---

[34] This expression, I have extendedly used in [40], is derived from Hutchins, who introduced the expression "mediating structure", which regards external tools and props that can be constructed to cognitively enhance the activity of navigating. Written texts are trivial examples of a cognitive "mediating structure" with clear cognitive purposes, so mathematical symbols, simulations, and diagrams, which often become "epistemic mediators", because related to the production of scientific results: "Language, cultural knowledge, mental models, arithmetic procedures, and rules of logic are all mediating structures too. So are traffic lights, supermarkets layouts, and the contexts we arrange for one another's behavior. Mediating structures can be embodied in artifacts, in ideas, in systems of social interactions [...]" ([31] pp. 290–291) that function as an enormous new source of information and knowledge.

non-cooperative (no teams are allowed), positions are extensible (that is they can be represented by board position trees).[35]

3.7.2   Locked Abductive Strategies Acting in a Fixed Scenario of Data Counteract the Maximization of Eco-Cognitive Openness

As I have already anticipated above in Sect. 3.1.2, in my research I have recently emphasized [47, Chapter seven] the *knowledge enhancing* character of abduction. In the case of an artificial game such as Go, the knowledge activated thanks to an intelligent choice of already available strategies or thanks to the invention of novel strategies and/or heuristics must also be considered a result of knowledge enhancing abduction.

I strongly contended that, to arrive to uberous selective or creative optimal abductive results, useful strategies must be applied, but it is also needed to be in presence of a cognitive environment marked by what I have called *optimization of eco-cognitive situatedness*, in which eco-cognitive openness is fundamental [45]. This feature of the cognitive environment is especially needed in the case of strong creative abduction, beyond a strict relationships with data, that is for example when the kind of novelty is not restricted to the case of the "simple" successful diagnosis. I said that, to favor good creative and selective abductive reasoning, strategies must not be "locked" in an external restricted eco-cognitive environment, such as in a scenario characterized by fixed defining rules and finite material aspects, which would function as cognitive mediators able to constrain agents' reasoning.

In summary, abductive processes to hypotheses—in a considerable quantity of cases, for example in science—are highly *information-sensitive*, and face with a flow of information and data uninterrupted and appropriately promoted and enhanced when needed (of course also thanks to artefacts of various kinds). This means that also from the psychological perspectives of the individuals the epistemological openness in which knowledge channeling has to be maximized is fundamental. It is the only way to create cognitive value both "in" and "out" data, as I have repeatedly said in the previous sections.

3.7.3   Locking Strategies Restricts Creativity

Optimization of situatedness is related to unlocked strategies. Locked strategies, such as the ones active in Go game, AlphaGo, AlphaZero, and other computational AI systems and deep learning devices, do not favor the optimization of situatedness. Indeed,

---

[35] Cf. Wikipedia entry Go (game) https://en.wikipedia.org/wiki/Go_(game) date of access 23 July 2022.

I have already contended above that, to obtain good creative and selective abductions, reasoning strategies must not be "locked" in bounded eco-cognitive surroundings (that is, in scenarios designed by fixed defining rules and finite material objects which would play the role of the so-called cognitive mediators). In this perspective, a poor scenario is certainly responsible for the minimization of the eco-cognitive openness and it is the structural consequence of the constitutive organization of the game Go (and also of Chess and other games), as I have already described in Sect. 3.7.1. I have said that in the game Go stones, board, and rules are rigid and so totally predetermined; what instead is undetermined are the strategies and connected heuristics that are adopted to defeat the adversary in their whole process of application.[36]

As I have already said, the available strategies and the adversary's ones are always *locked* in the fixed scenario: you cannot, during a Go game, play for few minutes Chess or adopt another rule or another unrelated cognitive process, affirming that that weird part of the game is still appropriate to the game you agreed to play. You cannot decide to change the environment at will so *unlocking* your strategic reasoning, for example because you think this will be an optimal way to defeat the adversary. Furthermore, your adversary cannot activate at his discretion a process of eco-cognitive transformation of that artificial game. On the contrary, in the example of scientific discovery, the scientist (or the community of scientists) frequently recur to disparate external models and change their reasoning strategies[37] to produce new analogies or to favor other cognitive useful procedures (prediction, simplification, confirmation, etc.) to enhance the abductive creative process.

The case of scenarios in human scientific discovery precisely represents the counterpart of the ones that are poor from the perspective of their eco-cognitive openness. Indeed, in these last cases, the reasoning strategies that can be endorsed (and also created for the first time), even if multiple and potentially infinite, are *locked* in a determined perspective where the components do not change (the stones can just diminish and put aside, the board does not change, etc.). I would say that in scenarios in which strategies are locked, in the sense I have explained, an *autoimmunization* [3, 52] is active, that constitutes the limitations that preclude the application of strategies that are not related to "pre-packaged" scenarios, strategies that would be foreigners to the ones that are strictly intertwined with the components of the given scenario. Remember I already said that these components play the role of *cognitive mediators*, which anchor and constrain the whole cognitive process of the game.

To summarize and further explain (by linking the problem of locked and unlocked strategies to the various cases of selective and creative abduction):

1. Contrarily to the case of high level "human" creative abductive inferences such as the ones expressed by scientific discovery or other examples of special exceptional intellectual results, the status of artificial games (and of their computational counterpart) is very poor from the point of view of the non-strategic knowledge involved. We are dealing with stones, a modest number of rules, and one board.

---

[36] Of course, many of the strategies of a good player are already mentally present thanks to the experience of several previous games.

[37] Many interesting examples can be found in the recent [53].

When the game progresses, the shape of the scenario is spectacularly modified but no unexpected cognitive mediators (objects) are appearing: for example, no diversely colored stones, or a strange hexagonal board. On the contrary, to continue with the example of high levels creative abductions in scientific discovery (for example, in empirical science), first of all the evidence is extremely rich and endowed with often unexpected novel features (not only due to modifications of aspects of the "same things", as in the case of artificial games). Secondly, the flux of knowledge at play is multifarious and is related to new analogies, thought experiments, models, imageries, mathematical structures, etc. that are rooted in heterogeneous disciplines and fields of intellectual research. In sum, in this exemplary case, we are facing with a real tendency to a status of optimal eco-cognitive situatedness.

2. What happens when we are dealing with selective abduction (for example in medical diagnosis)? First of all, evidence freely and richly arrives from several empirical sources in terms of body symptoms and data mediated by sophisticated artifacts (which also change and improve thanks to new technological inventions). Second, the encyclopedia of biomedical hypotheses in which selective abduction can work is instead locked,[38] but the reference to possible new knowledge (locally created of externally available) is not prohibited, so the diagnostic inferences can be enhanced thanks to scientific advancements at a first sight not considered. Third, novel inferential strategies and linked heuristics can be created and old ones used in new surprising ways but, what is important, strategies are not locked in a fixed scenario. In sum, the creativity that is occurring in the case of human selective abduction is poorer than the one active in scientific discovery, but richer than the one related to the activity of the locked reasoning strategies of the Go game and AlphaGo/AlphaZero, I have considered above.

3. In Go (and similar games) and in deep learning systems such as AlphaGo/AlphaZero, in which strategies and heuristics are "locked", these are exactly the only part of the game that can be improved and rendered more fertile: strategies and related heuristics can be used in a novel way and new ones can be invented. Anticipations as abductions (which incarnate the activities of "reading ahead") just affect the modifications and re-grouping of the same elements. No other types of knowledge will increase; all the rest remains stable.[39] Of course, this dominance of the strategies is the quintessence of Go, Chess, and other games, and also reflects the spectacularity of the more expert moves of the human champions. However, it has to be said that this dominance is also the reason that explains the fact the creativity at stake is even more modest than the one involved in the higher cases of selective abduction (diagnosis). This fact is also the reason that explains why the smart strategies of Go or Chess games can be more easily simulated, for example with respect to the inferences at play in scientific discovery, by recent artificial intelligence programs, such as the ones based on deep learning.

---

[38] It is necessary to select from pre-stored diagnostic hypotheses.

[39] Obviously, for example, new rules and new boards can be proposed, so realizing new types of game, but this chance does not jeopardize my argumentation.

The reader does not have to misunderstand me: I do not mean to minimize the relevance of creative heuristics as they work in Go and other board games. John Holland already clearly illustrated [26, 27] that board games such as checkers, as well as Go, are wonderful cases of "emerging" cognitive processes, where potentially infinite strategies favor exceptional games: even if simply thanks to a few rules regulating the moves of the pieces, games cannot be predicted starting from the initial configurations. While other cases of emerging cognitive processes (I have indicated the example of scientific discovery) characterize what can be called "vertical" creativity (that is, related to unlocked strategies), board games are examples of "horizontal" creativity: even if board games are circumscribed by locked strategies that constrain the game, "horizontal" creativity can show astonishing levels of creativity and skilfulness. We already said that these extraordinary humans skills have been notably appropriated by artificial intelligence software: the example given in this chapter is the one of AI deep learning heuristics that were able to *learn from* human games.

At the beginning of this subsection, I have illustrated some amazing performances of Google DeepMind's program AlphaGo/AlphaZero against human players and the fact the system showed to be able to "create" unconventional moves—never played by humans—thus building new strategies, a fact that undoubtedly favors the attribution to the system of actual "human" capacities, and we have to add, better than the ones of the more skilled humans. AlphaGo instructed itself to play "attending" (so to speak) thousands of games played by human beings thanks to "reinforcement learning", which refers to that activity of self-playing of the machine (the machine plays against itself) to further feed and adapt its own neural networks. In conclusion, even if based on what I called in this chapter locked strategies, and thus far from the highest levels of human creativity, AI deep learning system and various other programs can also offer chances for business and a good integration in the market.

3.8 Conclusion

Focusing on those particular kinds of abductive cognition that generate deep hypothetical results, I have illustrated how they are able to go beyond data exactly thanks to a rich cognitive intensity, in which asking "why" is the key-factor. To this aim I took advantage of the illustration of two kinds of abduction: (1) *model-based*, in which the cognitive activity of making intelligible the world, beyond what mere rough data indicates, exploits models, diagrams, and simulations of various nature, and (2) *manipulative*, by showing how we can find methods of manipulative constructivity, particularly able to potently overstep data. Auxiliary but not minor topics I have described regard two main additional issues: first of all the dangers that the current hyperbolized accent on the value of learning from data renders the role of human abductive "creative" cognition increasingly assaulted and jeopardized and, second, the challenges against human abduction and epistemic rigor on the part of what I call computational invasive "subcultures" regarding big data. I devoted the last part of the chapter, thanks to the analysis of the famous computational deep learning program AlphaGo/AlphaZero, to the description of the important distinction between *locked* and *unlocked* strategies, which is stringently implicated in the generation of different cognitive outputs/hypotheses, from the weak (closer to data) to the strong ones (beyond data), characterized by higher levels of knowledge creativity. Locked abductive strategies, which in this perspective appear cognitively weak, paradoxically characterize (sometimes amazing) types of *hypothetical creative reasoning* and are distinctive of intelligent machines, but they are bounded in *eco-cognitive openness*, that instead regards human cognizers who are executing higher kinds of abductive creative reasoning, in which cognitive strategies are unlocked.

Acknowledgements Parts of this chapter are excerpted from chapters five, six, and eight of my book *The Abductive Structure of Scientific Creativity. An Essay on the Ecology of Cognition*, Springer, Cham, 2017 and from my article AlphaGo, locked strategies, and eco-cognitive openness. *Philosophies*, 2019, 4(1):8. For the instructive criticisms and precedent discussions and correspondence that helped me to develop my analysis of abductive cognition I am indebted and grateful to John Woods, Atocha Aliseda, Woosuk Park, Yukio Ohsawa, Luís Moniz Pereira, Paul Thagard, Ping Li, Gerhard Schurz, Walter Carnielli, Akinori Abe, Shahid Rahman, Gerhard Minnameier, and to my collaborators Selene Arfini and Alger Sans Pinillos.

References

1. Aliseda, A.: Seeking explanations: abduction in logic, philosophy of science and artificial intelligence. Ph.D. Thesis, Institute for Logic, Language and Computation, Amsterdam (1997)
2. Aliseda, A.: Abductive Reasoning. Logical Investigations into Discovery and Explanation. Springer, Heidelberg/Berlin (2006)
3. Arfini, S., Magnani, L.: An eco-cognitive model of ignorance immunization. In: Magnani, L., Li, P., Park, W. (eds.) Philosophy and Cognitive Science II. Western & Eastern Studies, vol. 20, pp 59–75. Springer, Switzerland (2015)
4. Bacon, F.: The New Organon [1620]. Cambridge University Press, Cambridge (2000)
5. Biot, J.B.: On the magnetism impressed on metals by electricity in motion. Q. J. Sci. **11**, 281–290 (1821) . Read at the public setting of the Academy of Sciences, 2nd April, 1821
6. Brown, B., Priest, G.: Chunk and permeate, a paraconsistent inference strategy. Part I: the infinitesimal calculus. J. Philos. Logic 33(4), 379–388 (2004)
7. Calude, C.S., Longo, G.: The deluge of spurious correlations in big data. Found. Sci. **22**(3), 595–612 (2017)
8. Carnielli, W.: Surviving abduction. Logic J. IGPL **14**(2), 237–256 (2006)
9. Coelho, H., Thompsen Primo, T.: Exploratory apprenticeship in the digital age with AI tools. Progr. Artif. Intell. **1**(6), 17–25 (2017)
10. D'Avanzo, E., Zhuhadar, L., Lytras, M.D. (eds.): Dig data research and internet of things research: a new digital vision for the knowledge society. Special Issue of the Int. J. Knowl. Soci. Res. **3**(2) (2016)
11. Davies, J.: Tesuji. Elementary Go Series 3. Kiseido Publishing Company, Tokyo (1995)
12. Davy, H.: On the magnetic phenomena produced by electricity. Philos. Trans. **111**, 7–19 (1821)
13. Einstein, A.: Relativity and the problem of space [1952]. In: Ideas and Opinions, pp. 360–377. Crown Publisher, New York, (2014). Translated by S. Bergmann
14. Estrada-González, L.: Remarks on some general features of abduction. J. Log. Comput. **232**(1), 181–197 (2013)
15. Fan, J., Han, F., Liu, H.: Challenges of big data analysis. Int. J. Inf. Manage. **1**(2), 293–314 (2014)

16. Faraday, M.: Historical sketch on electromagnetism. Ann. Philos. **18**, 195–200, 274–290; **19**, 107–121 (1821–1822)
17. Feyerabend, P.: Against Method. Verso, London-New York (1975)
18. Franklin, I.R.: Exploratory experiments. Philos. Sci. **72**, 888–899 (2005)
19. Frizzo-Barker, J., Chow-White, P.A., Charters, A., et al.: Genomic big data and privacy: challenges and opportunities for precision medicine. Comput Supported Coop Work **25**(2–3), 115–136 (2016)
20. Gabbay, D.M., Woods, J.: The Reach of Abduction. North-Holland, Amsterdam (2005)
21. Gandomi, A., Haider, M.: Beyond the hype: big data concepts, methods and analytics. Int. J. Inf. Manage. **35**(2), 137–144 (2015)
22. Gibson, A., Bruza, P.: Transepistemic abduction: Reasoning across epistemic domains. Logic J. IGPL **29**(4), 469–482 (2021). Special Issue on "Formal Representations in Model-Based Reasoning and Abduction", ed. by A. Nepomuceno, L. Magnani, F. Salguero, Barés-Gómez, M. Fontaine
23. Gooding, D.: Experiment and the Making of Meaning. Kluwer, Dordrecht (1990)
24. Gooding, D., Addis, T.R.: Modelling experiments as mediating models. Found. Sci. **13**(1), 17–35 (2008). Special issue edited by L. Magnani "Tracking irrational sets. Science, technology, ethics", Proceedings of the International Conference "Model-Based Reasoning in Science and Engineering" - MBR04
25. Hendricks, F.V., Faye, J.: Abducting explanation. In: Magnani, L., Nersessian, N.J., Thagard, P. (eds.) Model-Based Reasoning in Scientific Discovery, pp. 271–294. Kluwer Academic/Plenum Publishers, New York (1999)
26. Holland, J.H.: Hidden Order. Addison-Wesley, Reading, MA (1995)
27. Holland, J.H.: Emergence: From Chaos to Order. Oxford University Press, Oxford (1997)
28. Horsman, C., Stepney, S., Wagner, R.C., et al.: When does a physical system compute? Proc. R. Soc. A **470**, 1–25, 470 20140182 (2014). https://doi.org/10.1098/rspa.2014.0182. Accessed 9 July 2014
29. Husserl, E.: Ideas. General Introduction to Pure Phenomenology [First book, 1913]. Northwestern University Press, London and New York (1931). Translated by W. R. Boyce Gibson
30. Husserl, E.: The Origin of Geometry (1939). In: Derrida, J. (ed.) Edmund Husserl's "The Origin of Geometry", pp. 157–180. Nicolas Hays, Stony Brooks, NY (1978). Translated by D. Carr and originally published in [Husserl E (1970) The Crisis of European Sciences and Transcendental Phenomenology [1954]. George Allen & Unwin and Humanities Press, London and New York, translated by. D. Carr], pp. 353–378
31. Hutchins, E.: Cognition in the Wild. The MIT Press, Cambridge, MA (1995)
32. Kakas, A., Kowalski, R.A., Toni, F.: Abductive logic programming. J. Log. Comput. **2**(6), 719–770 (1993)
33. Kowalski, R.A.: Logic for Problem Solving. Elsevier, New York (1979)
34. Kuipers, T.A.F.: Abduction aiming at empirical progress of even truth approximation leading to a challenge for computational modelling. Found. Sci. **4**, 307–323 (1999)
35. Livio, M.: Brilliant Blunders: From Darwin to Einstein. Colossal Mistakes by Great Scientists That Changed Our Understanding of Life and the Universe. Simon & Schuster, New York, NY (2013)
36. Longo, G.: Critique of computational reason in the natural sciences. In: Gelenbe, E., Kahane, J.P. (eds.) Fundamental Concepts in Computer Science. Imperial College Press/World Scientific, London (2009)
37. Longo, G.: Incomputability in physics and biology. Math. Struct. Comput. Sci. **22**(5), 880–900 (2012)
38. Longo, G.: The biological consequences of the computational world: mathematical reflections on cancer biology (2017). arXiv:1701.08085
39. Longo, G., Tendero, P.E.: The differential method and the causal incompleteness of programming theory in molecular biology. Found. Sci. **12**(4), 337–366 (2007)
40. Magnani, L.: Abduction, Reason, and Science. Processes of Discovery and Explanation. Kluwer Academic/Plenum Publishers, New York (2001)

41. Magnani, L.: Abductive Cognition. The Epistemological and Eco-Cognitive Dimensions of Hypothetical Reasoning. Springer, Heidelberg/Berlin (2009)
42. Magnani, L.: Is abduction ignorance-preserving? Conventions, models, and fictions in science. Logic J. IGPL **21**(6), 882–914 (2013)
43. Magnani, L.: The eco-cognitive model of abduction. ᾽Απαγωγή now: Naturalizing the logic of abduction. J. Appl. Logic **13**, 285–315 (2015)
44. Magnani, L.: Naturalizing logic. Errors of reasoning vindicated: logic reapproaches cognitive science. J. Appl. Logic **13**, 13–36 (2015)
45. Magnani, L.: The eco-cognitive model of abduction. Irrelevance and implausibility exculpated. Journal of Applied Logic **15**, 94–129 (2016)
46. Magnani, L.: Playing with anticipations as abductions. Strategic reasoning in an eco-cognitive perspective. J. Appl. Logic – IfColog J. Logics Appl. **5**(5), 1061–1092 (2017). Special issue "Logical Foundations of Strategic Reasoning", edited by W. Park
47. Magnani, L.: The Abductive Structure of Scientific Creativity. An Essay on the Ecology of Cognition. Springer, Cham, Switzerland (2017)
48. Magnani, L.: Eco-cognitive computationalism: From mimetic minds to morphology-based enhancement of mimetic bodies. Entropy **20**(6) (2018)
49. Magnani, L.: AlphaGo, locked strategies, and eco-cognitive openness. Philosophies **4**(1), 8 (2019)
50. Magnani, L.: Abduction as "leading away". Aristotle, Peirce, and the importance of eco-cognitive openness and situatedness. In: Shook, J., Paavola, S. (eds.) Abduction in Cognition and Action: Logical Reasoning, Scientific Inquiry, and Social Practice, pp. 77–105. Springer, Cham (2021)
51. Magnani, L.: Computational domestication of ignorant entities. Unconventional cognitive embodiments. Synthese **198**, 7503–7532 (2021). Special Issue on "Knowing the Unknown" (guest editors L. Magnani and S. Arfini)
52. Magnani, L., Bertolotti, T.: Cognitive bubbles and firewalls: Epistemic immunizations in human reasoning. In: Carlson, L., Hölscher, C., Shipley, T. (eds.) CogSci 2011, XXXIII Annual Conference of the Cognitive Science Society. Cognitive Science Society, Boston MA (2011)
53. Magnani, L., Bertolotti, T. (eds.): Handbook of Model-Based Science. Springer, Switzerland (2017)
54. Meheus, J., Verhoeven, L., Van Dyck, M., et al.: Ampliative adaptive logics and the foundation of logic-based approaches to abduction. In: Magnani, L., Nersessian, N.J., Pizzi, C. (eds.) Logical and Computational Aspects of Model-Based Reasoning, pp. 39–71. Kluwer Academic Publishers, Dordrecht (2002)
55. Minski, M.: The Society of Mind. Simon and Schuster, New York (1985)
56. Park, W.: On classifying abduction. J. Appl. Log. **13**, 215–238 (2015)
57. Peirce, C.S.: Collected Papers of Charles Sanders Peirce. Harvard University Press, Cambridge, MA (1866–1913), vols. 1–6, Hartshorne, C., Weiss, P. (eds.); vols. 7–8, Burks, A. W. (ed.) (1931-1958)
58. Port, R.F., van Gelder, T. (eds.): Mind as Motion. Explorations in the Dynamics of Cognition. The MIT Press, Cambridge, MA (1995)
59. Sax, M.: Big data: finders keepers, losers weepers? Ethics Inf. Technol. **18**(1), 25–31 (2016)
60. Thagard, P., Shelley, C.P.: Abductive reasoning: logic, visual thinking, and coherence. In: Chiara, M.L.D., Doets, K., Mundici, D., et al. (eds.) Logic and Scientific Methods, pp. 413–427. Kluwer, Dordrecht (1997)
61. Tweney, R.D.: Abductive seeing. Unpublished Manuscript (2006)
62. Woods, J.: Errors of Reasoning. Naturalizing the Logic of Inference. College Publications, London (2013)
63. Woods, J.: Reorienting the logic of abduction. In: Magnani, L., Bertolotti, T. (eds.) Springer Handbook of Model-Based Science, pp. 137–150. Springer, Cham (2017)

Chapter 4
Living Labs: A Device That Opens Exploration and Cognitive Generation to Society

Sae Kondo and Yukio Ohsawa

Abstract Here, we argue that the full value of data in society cannot be taken without communication about the real life and thoughts of people, which are the latent dynamics behind data. As an embodiment of this approach, we discuss the method, role, and potential advantages of living labs, in which participants from different walks of life come together to share their issues, perspectives, and wisdom to work out solutions. Reviewing the global history of the development of Living Labs, the types, scope, and methods of activities, as well as case studies, it was clarified that Living Labs provide steps of exploration that open up different perspectives to each other. Through these steps, participants discovered the essential problems that they felt they had to solve. This chapter is positioned here as a guide for the reader to widen the scope of using AI with existing data in one's machine to data collected in the future of the people used by those people for those people themselves to live beyond the data.

Keywords Living Labs · Citizen participatory design · Open data · Open government · ENoLL · Smart city

4.1 Introduction

Urban planning in the 20th century has been developed from citizens' point of view, intending to improve their quality of life. However, only a few citizens (of certain social status or position) expressed their views. Although bottom-up urban planning has become more widespread, the number of people who speak up and take action is limited. Moreover, in the 21st century, when the speed and variety of ways of mobility

S. Kondo (✉)
Department of Architecture, Graduate School of Engineering, Mie University, Mie, Japan
e-mail: skondo@arch.mie-u.ac.jp

Y. Ohsawa
Department of Systems Innovation, School of Engineering, The University of Tokyo, Tokyo, Japan
e-mail: ohsawa@sys.t.u-tokyo.ac.jp

and information exchange have increased dramatically, 'quality' of life is perceived to be uniform wherever you go, as long as you have the latest devices at hand in a city with easy access to the Internet. In this sense, young people, in particular, are less likely to find value in the space and environment in which they live. On the other hand, politicians and administrators responsible for improving the urban environment must explain that the measures they provide reflect citizens' views, because citizens' taxes fund them. In response to this dilemma, there has been growing interest in EBPM, the use of data in decision-making, such as tracking people's behaviour, as an alternative to the voice of the citizen. However, data represent only observable phenomena, for example, regarding each movement of the hands and feet in human behaviour, and it is not easy to understand deeper than the physical behavioural level. Although deep learning has enabled the realisation of human-life cognitive mechanisms to dig deep into latent factors in real-world targets [4, 7, 13], the will and intentions that determine the behaviour of individuals and people as a society or the demands to satisfy their daily requirements are difficult to explain from the analysis of data, for example, behavioural logs, text in SNS, or items in POS data. This kind of explanation is essential in practical business and for the satisfaction of living people, as shown in other chapters of this book. It is only by exploring and interpreting the latent dynamics behind data from the broader perspective of society that we can acquire an understanding of this explanation, which means grasping the full picture of the underlying meaning and shaping its use. It is up to us, as human beings live in diverse and complex societies, to grasp this picture. The ability of humans to grasp a picture has been studied as an extension of sense-making and chance discovery [6, 9]. To accelerate the exploration and interpretation of meaning, this chapter suggests that it is necessary to have a mechanism for people from various backgrounds to attend the Living Lab, a place where citizens from all walks of life can come together to share their perspectives and knowledge of the facts and issues that exist in front of them and work to improve and solve them. The questions we show and challenge here are: What kinds of experiences do Living Labs provide to the people who gather? What do they acquire from these experiences? We want to discuss the roles and possibilities of Living Labs.

4.2 Overview of European "Living Labs"

4.2.1 Living Lab's Origins and Founding Intentions

According to the white paper No.4 of the Nordic Institute, an innovation method called "participatory design", which involves citizens and other stakeholders and is implemented by the whole community has been proposed in northern Europe since the 1970s. In recent years, succeeding the method proposed here, "Living Labs" have been attracting attention as an effective and sustainable approach to innovation for challenging the increasingly complex and uncertain society. Living Labs and "participatory design" are terms and approaches that are widely used in North America and Europe. Whereas the North American approach to community building is based on leadership debates faced toward the market, the Scandinavian Living Labs and participatory design adopt a more holistic, socially inclusive, and democratic approach using a dialogue method to find a middle ground. The Scandinavian Living Labs are regarded as one way for those who want to listen to the voices and share the intentions of people to be reflected in the improvement of their lives and the urban environment.

4.2.2 Development History of Living Labs

In this section, we provide an overview of the development of Living Labs in Europe.

4.2.2.1 Development History/Short History

There are three predecessors to the development of Living Labs in Europe [10].

The first was the social movement of cooperative and participatory design, which flourished in Scandinavia during the 1960s and 70s. Many research projects throughout Scandinavia introduced user participation in the system development during this period. However, cooperative design realised that further technical development could not be achieved by designing in a closed way within an already established community, and the so-called user-centered design approach emerged. This method and participatory design are still widely used, particularly in Scandinavian Living Labs.

Second, many social experiments with IT were launched across Europe in the 1980s, partly funded by the European Commission, and researchers in ICT adopted social experiments to demonstrate and implement technologies. For example, France, Germany, the UK, and Denmark participated in an interactive videotex. This period also marked the beginning of the EU's interest in the growth of Europe through the opening and use of public sector data and the beginning of the convergence of public and private expectations in the development of data utilisation technologies [8].

The third is the concept of a digital city, adopted in Europe and elsewhere in the 1990s. This refers to several digital initiatives undertaken by cities, such as digital-related economic development, urban regeneration, and Internet access for citizens. It links citizens (users), policymakers (public authorities), and private organisations (businesses) on a large scale. The topics covered here were diverse but always linked to urban life. Then, it became considered that citizens were considered potential creative agents for innovation, the technological tools and infrastructures introduced were considered triggers for creativity, and the user involvement approach was inevitably assumed to be multi-stakeholder. This can be understood through theoretical and experimental studies [15].

As a corollary, in the early 1990s in the United States, as the Internet opened up to individual users, governmental data began to be released online (starting with the public posting of California State legislative records on the Internet). It was not just a development of technology to improve accessibility to public data but also led to the development of the 21st-century theory of "open government", which aims to give meaning to open data and increase public trust and national development [16]. Thus, the easy and low-cost availability of public sector data at the US federal level and the drive to promote the use of data in the public and private sectors posed a threat to the underdeveloped and uncompetitive European information market [8]. In 2003, the EU adopted the Public Sector Information (PSI) Directive [1]. It was recognised that the larger availability of public sector data and value-added information products and services would benefit the entire society, then the main target group of the directive was the information industry, i.e., "the creation of Community-wide information products and services based on public sector documents". In addition to the three forerunners mentioned above, it must be added that the term "Living Labs" has been used incidentally for some time. It is said that the concept originated by William Mitchell of MIT, who used it to mean "a dedicated laboratory in which the activities and interactions of everyday domestic life can be observed, recorded, analysed and experimented with, and in which volunteer researchers live individually as temporary residents"; The MIT Place Lab is an example of this. Mitchell became a member of the Expert Advisory Group of the "Intelligent Cities—Intelcities" project (2002–2005), which involved European cities, and his work at MIT became well known in Europe. On 20 November 2006 during the presidency of the European Union, the President of Finland launched the European Living Lab Network (ENoLL) with the cities involved in the project as the first step toward a new European innovation system based on co-creative innovations through partnerships between the public, private, and citizens. The ENoLL now acts as a platform for information-sharing and support for living labs worldwide.

The 2000s were a period of growing momentum for Living Labs owing to a series of EU policy measures. Two of the earliest policies in 2006 were the "Corelabs" and "Clock" projects. They aimed to roll out the ICT innovation system across Europe through living labs. Nineteen Living Labs from across Europe have been incorporated. Most of these were partners in the "Intelcities" project, which ran until the year before the "Corelabs" and "Clock" projects began. In addition, the EC's Information Society and Media Directorate, New Collaborative Working Environments Unit, has allocated 40 million euros to the activities of the European Living Lab Network since the fifth call for proposals for the Sixth Framework Program. The Living Labs then became an opportunity to bridge the gap between European policies and programmes such as Horizon 2020, Smart Specialisation, Urban Agenda, and Cohesion policy, grounded in the needs and aspirations of local and regional stakeholders, and it was an opportunity to realise a bottom-up contribution to policymaking. This trend was particularly evident in policies and practices around the "smart city" concept, such as the FIREBALL (Future Internet Research and Experimentation By Adopting Living Labs) project. For example, the FIREBALL project has led to the development of strategic policies such as the Smart City Roadmap and the White Paper through

Living Labs. In addition, ENoLL currently has more than 150 active members world-wide, working as an information-sharing and co-creative network for innovations in various sectors such as energy, media, mobility, healthcare, and agri-food [2]. It aims to act as a platform for exchanging, learning, and supporting best practices in user engagement, testing, and placemaking as well as for the development of Living Labs' international projects.

4.2.3 Types of Living Labs, Scope of Activities and Methods

In this section, we introduce the types of Living Labs and their activities. We also mention, in a limited way, the methodology of Living Labs.

4.2.3.1 Types of Living Labs

ENoLL, which collects information on many living labs, classifies their members into the following four types of Living Labs based on key principles, key components, and the resulting values shown in Table 4.1 (Ståhlbröst 2012). It should be noted that not all living labs have to be one of these four types and that combinations of these and new types of living labs can exist.

(1) Research Living Labs focus on performing research on different aspects of the innovation process.
(2) Corporate Living Labs that focus on having a physical place where they invite other stakeholders (e.g. citizens, collaborators, resource suppliers) to co-create innovations.
(3) Organizational Living Labs, where members of an organisation co-creatively develop innovation.
(4) Intermediary Living Labs, in which partners from different domains are invited to innovate collaboratively in a neutral arena.

4.2.3.2 Scope of Activity

As mentioned above, there are different types of living labs, and the ENoLL acknowl-edges that living labs use their project methods according to their objectives and actors. However, Living Labs have developed research projects aimed at systems innovation in IT and ICT and their penetration into everyday life. The range of applications from this perspective is presented in Table 4.2 [14]. This table shows that the Living Labs are diverse, dealing with a wide range of tools that can bring richness and convenience to many aspects of our lives, which is not surprising since IT and ICT technological development cannot be grouped under a single academic discipline. However, what is consistent is that 'user behaviour' is a measure of the

Table 4.1 Living Labs key 5 principles and key components by ENoLL [12]

5 key principles	Key components
• Value: should support value creation in preferably two different ways; business value for partners and user or social value for stakeholders of innovation • Influence: As user involvement and influence is essential to the innovation process, we view users as active and competent partners and domain experts and encourage their participation and involvement by explaining the impact of their interaction on innovation • Realism: One of the cornerstones of Living Labs is the understanding that activities should be carried out in a realistic, natural, real-life setting. This setting comes from the belief that people cannot experience anything independent of the their embodied life in the world • Sustainability: Living Labs is defined as an approach that meets the need of the present without compromising the ability for future generations from an economic, social and ecological perspective. It is also one cornerstone for the continuous learning cycle within Living Labs • Openness: This principle stresses the importance of having an innovation process that supports a bidirectional flow of knowledge and resources between stakeholders [5]. The idea is that multiple perspectives empower the process and contribute to the achievement of rapid progress. However, to be able to cooperate and share in a multi-stakeholder milieu, different levels of openness between stakeholders come to be a requirement	• ICT • Infrastructure • Management • Partners and Users • Research and Approach

effectiveness of developed products, technologies, and services. Nevertheless, it is common for companies that manufacture products and services to carry out usability testing before bringing them to the market, so what makes Living Labs different from the rest? It can be said that what makes Living Labs different is that participants go beyond usability tests and consider the timeline of life and the background of the consumer as important factors in deepening and broadening their interpretations. If they want to bring more innovation to the future of overabundant lives, participants need to consider the perspectives and values of the people who use the developed products, as well as those who make them. The Living Lab's Key Principles of "Value", "Influence", and "Realism" make this clear.

4.2.3.3 Living Labs Methodology

We have already mentioned that each Living Lab uses its project methodology depending on its objectives and participants, who should be referred to as actors, considering the activity-oriented nature of Living Labs. However, let us give Living Labs slightly more apparent contours, but in a limited manner. Overall, the activities

Table 4.2 The scope of Living Labs for sustainable development/Rohn and Leismann, in [14]

Fields of application	Research perspectives	
	Product and service innovation	User behaviour
Living and working environment	Buildings and infrastructure: security, heat and energy supply, insulation, e-energy assistant Food: refrigeration and freezing techniques, cooking, food assistant Residential and office settings: appliance design, furniture and textiles Information management Physical mobility: logistics systems based on "ICT mobility", smart grids	Residential and office behaviors: health, exercise and energy consumption Nutrition: food waste, shopping, health Designing homes and offices for different phases of life: autonomous living in old age, user acceptance of innovations Integrated design: demand and service sectors Housing and office set-up: new workplace design, ICT Services and time management: healthy eating, exercise
City, region and transport	Catering: delivery service, drive-in restaurants Mobility: optimal logistics, aviation, public transport links, design of mobility options Regional networks, regional PR: health support systems, urban planning, communication systems, regional energy supply, tourism, sharing and rental	Transport: resource-efficient mobility options and user acceptance Information networks: urban agriculture, barter systems, neighborhood networks, service concepts and suburban development Leisure activities: regional tourism ICT services: transport and logistics management
Retail, food culture	Set-up/furnishing: appliances, lighting, media, online shopping, design Transport: an efficient way to get around Nutrition: food labelling and declaration Old age support: intelligent household appliances	Intelligent Consumer Electronics: Digital Product Memory Product choice: the impact of advertising and information campaigns

in Living Labs are designed to create something new that will improve the quality of our lives. In other words, it was a creative activity. This process consists of three phases [10]. It follows the experience gained by the three predecessors in developing Living Labs.

The three phases of the Living Labs procedure:
(Phase 1) Users participate in the exploration process corresponding to the collaborative and participatory design in the first predecessor stage.

(Phase 2) Creative ideas from exploration are turned into prototypes and tested, corresponding to social experiments in the second predecessor stage.
(Phase 3) The project results are commercialised and implemented in society through policymaking, corresponding to the realisation of a digital city.

Thus, from this component of the process, Living Labs fulfil the function of a process generator of creativity. However, bringing diverse people into the creative process is not uniform and it is challenging to present the best and most complete versions. The psychologist Finke, who presented the "Geneplore model" as a useful framework for describing the basic cognitive processes associated with creativity, also understands that there are many different aspects of the creative act, and that it is more appropriate to view these in a structured way and interpret them in a meaningful way [11]. This means that the most important thing for Living Labs is not to obtain a general solution to its method but to understand that there are different levels of granularity in the process (as mentioned above, the process has three phases, each with different themes, participants, goals, etc.) and the hierarchy between them. An appropriate combination of these factors should allow participants to be creative and ultimately achieve their goals.

4.3   The Case of "Living Labs"

4.3.1   Botnia Living Lab—for Sustainable Smart Cities and Regions

This active and long-lasting Living Lab began in 2000, mainly at the Lulea University of Technology in Sweden, funded by the European Commission. It advocates a world-leading environment for innovation through user-centered research and development, and is equipped with experts. It is also a member of the ENoLL and has inspired many Living Labs. Botnia provides methods and tools for interacting with user groups (Fig. 4.1). The methodology, named FormIT, is described as different from the problem-solving approach, and encourages users to iterate through the three actions of requirement generation, design, and evaluation in all design phases: concept, prototype, and innovation (Fig. 4.2). This envisages a spiral process that connects design phases.

4.3.2   HSBC Living Lab, Chalmers University of Technology

The HSBC Living Lab is an experimental housing project located at Chalmers University of Technology in Gothenburg, Sweden, developed by the HSB Housing Association, Chalmers University of Technology, and Gothenburg Science Park. It was part of the SubLabNWE project, implemented in four locations in Sweden, Ger-

Fig. 4.1 The handbook of Botnia Living Lab https://www.ltu.se/research/subjects/information-systems/Botnia-Living-Lab/Handbocker?l=en (accessed 2022-01-29)

many, and the UK from 2012 to 2015 with EU regional development funding. It has been operating independently since the completion of the project. This project aimed to experimentally test technologies specific to sustainable lifestyles, such as energy efficiency and energy consumption in housing, in a simulated living environment (see Figs. 4.3 and 4.4). The project is a co-creation of industry, government, and academia, and consists of a hierarchical three-layered management structure: a group that considers the development and operational issues of the project, a group that makes decisions, and an owner's group that solves problems. HSB's focus has now evolved to open collaborative projects, such as urban development. Each partner from industry, government, and academia has a direct contract with the HSB and is on an equal footing, defining together the relationship with the Living Lab, intellectual property issues, sources of income, etc. (a consortium contract between the partners is under consideration). In addition to the HSB Living Lab projects, several other exciting projects add value to the partners' arena of activity and contribute directly to progress [3]. The HSB approach, in which partners donate money and time, may be one way to support the sustainability of Living Labs.

4.3.3 Trends in Japan

We have described the background of interest in Living Labs in Japan in the Introduction section. To see how interest has grown, let us look at articles about Living Labs published in four major Japanese newspapers (Asahi, Yomiuri, Nikkei, and Mainichi) in the past 14 years from 2019 (Table 4.3). According to this table, in the past few years, living labs in Japan have tended to be taken up as a co-creative mechanism and organisation for solving social and regional problems rather than as

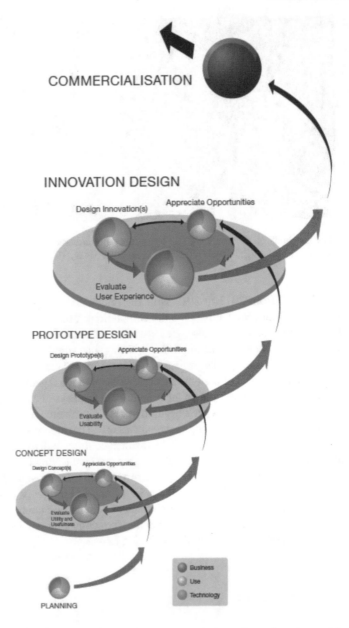

Fig. 4.2 The FormIT process by Botnia Living Lab https://www.ltu.se/research/subjects/information-systems/Botnia-Living-Lab/Handbocker?l=en (accessed 2022-1-29)

Fig. 4.3 HSB Living Lab: the residential building which is installed experiment equipment

Fig. 4.4 Inside of HSB Living Lab

a co-creative activity for developing innovations in industry. In other words, it has only recently that Living Labs have been recognised as a device for open innovation to solve social problems. This is probably because information about Living Labs in Europe has gradually been introduced to Japan piecemeal.

Table 4.3 Content of newspaper articles about the Living Labs from 2006 to 2019

Year	Name of LL	Features of the activities at LL
2006–2012	• Living Lab (TOTO) • Living Lab • Lions Living Lab	Activities by the company's in-house validation studio, home equipment showroom and women's product development team
2013–2016	• Living Lab Tokyo • Full board living lab • Shinshu Healthcare Equipment Living Lab • Faculty-led laboratories aiming to develop new technologies and devices close to everyday life • Citizens are invited to use the products and prototypes, and the company collects data to develop products	• Faculty-led laboratories aiming to develop new technologies and devices close to everyday life • Citizens are invited to use the products and prototypes, and the company collects data to develop products
2016–2019	• Wise Living Lab • Kamakura Living Lab • Matsumoto Health Lab • Hiranuma Living Lab • Rokutukawa Living Lab • SDGs Yokohama-Fukuzawa Living Lab etc.	• Public consultations to obtain opinions from within and outside the town and reflect them in future planning • Aiming to solve regional issues such as disaster prevention and transportation through "collaboration between industry, academia, public and private sectors" by holding seminars etc. • In an ageing residential area, universities, civil society and the government are working together to solve local (regional) problems • Stimulation of the local economy

4.4 The Essential Features of a Living Labs

As shown above, living laboratories use their methods for projects according to their objectives and actors, and their methods are not uniform in all situations. However, they all share the same approach, which is to generate ways to improve and enhance the quality of social life by externalising the fundamental problems hidden in the everyday experiences of diverse people via the bottom-up interaction of participants and wider society. Finke explained that the exploration phase (corresponding to the interpretation of the meaning of a produced product, service, or system) follows the generation phase (corresponding to production). However, since Living Labs are significant in gathering the perceptions of a wide range of people, starting from the fact that one person's perspective captures only a small part of society, it begins with the exploration phase in which we borrow the perspectives of others to re-interpret the events at hand. In this sense, it can be said that the process follows a diffusive exploration phase, a pre-production phase, and a convergent search phase followed by the a production phase. However, what is important is not that we proceed in this order precisely but that we are flexible enough to allow different perspectives to open up to each other, connect and create procedures as appropriate, and bring many people together. The more diverse the group of people, the more the exploration process allows for the recognition of real problems hidden behind the data, and the more it allows for the union and synthesis of cognitions that are meaningful to both oneself and society. Living Labs provide participants with steps of exploration, in which different perspectives open up to each other. Through these steps, participants could open their minds and identify the fundamental problems that they felt they needed to solve. In the next section, we present the activities of Living Labs in light of this essential function.

4.5   A Case Study of Living Labs Based on Its Essential Functions in Iwaki City, Japan

The city of Iwaki in Fukushima Prefecture, a nuclear power plant located in a neighbouring municipality, was hit by a tsunami during the Great East Japan Earthquake on 11 March 2011. Since then, the city has been working toward the industrialisation of renewable energy with support from the national government. However, to contribute to the sustainability of the region, it is essential that not only experts and central funding are involved, but that businesses and even individual citizens who are not directly involved in the industrial sector should share the vision for the future of the region as a whole to think and act proactively in developing the region. Therefore, it is desirable to develop a basis for open innovation. To achieve this, Living Labs were set up to encourage various businesses in the city to reinterpret their activities from an environmental perspective, and to search for and implement business activities that could also contribute to the local community.

[Contents of the Iwaki City Living Lab]
The Iwaki Living Lab initiative consisted of workshops on the topics listed in Table 4.4, and energy sector experts visited each participating business partner to provide consultancy on eco-action (conducted on various occasions).
[Workshop Position]
The experts provided information on predefined themes during the first half of each workshop. In the second half, the activities of each company were reinterpreted from the perspective of the theme to uncover latent issues, and the participants shared their wisdom to find solutions to these issues. We arranged a diffusive exploration and preproduction set, which was repeated for the first five workshops. In the final session, we proceeded with a convergent exploration phase. We shared our vision of a sustainable region, considering ways to solve the externalised individual and regional problems, and proposed actions that each company could take in this context. As mentioned in the previous section, the order of each step in the procedure is not essential.

Table 4.4 Iwaki Living Lab's activities

Date	Workshop theme	Aim
2019.09	1. How to engage in eco-action	Consider the challenges, resources required and solutions for implementing eco-actions in one's belonging company that do not lead directly to profit
2019.11	3. Visualisation of the flow of all goods	Review your company's work flow, identify bottlenecks and think about solutions
2019.12	4. Identifying resource wastage	Thinking about why waste is a problem for one's belonging company and how it can be put to good use
2019.12	5. Thinking about the process of making a medium- to long-term business successful	Consider the business processes and business development that need to be organised from the start-up to the start of operation, including internal consensus building, when implementing a medium- to long-term project such as wind power on one's own
2020.01	6. Challenging our own and local issues "Iwaki City Future Vision"	Based on the learning and discussions up to the 5th session, each company will set up a "vision for the future of Iwaki City" and create an action plan

Table 4.5 Iwaki Living Lab's participants

No.	Industry	Number of participants
A	Chemical industry	1 (Managerial level)
B	Road freight forwarding	1 (executive level)
C	Food and Beverage Retailing	1 (Managerial level)
D	Non-ferrous metal manufacturing	3 (Section Chief)
E	Manufacture of electrical machinery and equipment	2 (general manager and section manager level)
F	Information services	1 (executive level)

[Participant]

As a prerequisite that the participants would consist of different industries and fields, we selected businesses in the city belonging to the Iwaki City Chamber of Commerce and Industry and invited them to participate. Nine people from various industries and business sections participated in the study, as shown in Table 4.5.

[Project achievement]

In all the "visions for the future of Iwaki City" presented by the participants in the last session, their company's new or further contribution to the community was drawn up. In particular, the proposal by Company C and Company F is to use surplus food products from Company C for the welfare and healthcare needs of night shift workers from other participating companies. The business model aimed to provide a stable supply of agricultural products to the region and improve the food security of the

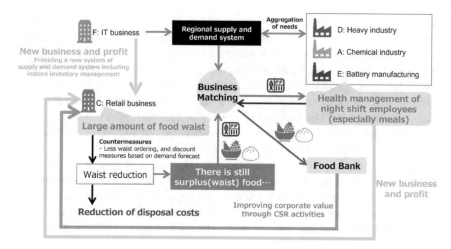

Fig. 4.5   Results of the Iwaki LL/Circular business model for local resources

elderly and children in the region (Fig. 4.5). The remaining participants agreed to this proposal and offered their assistance. In the Iwaki Living Lab, participants discovered the critical problems they needed to tackle through socially open exploration and thought of solutions for their profit and the whole community's benefit. In addition, by bringing together businesses from different industries that had not previously interacted, they were able to generate and use unused and new data, thus achieving the objective of this initiative. We can say that the essential function of Living Labs was manifested. In the interviews conducted with the participants after completing the project, they felt that the connection and exchange with other industries broadened their horizons. They also pointed out that they had not had the opportunity to interact with other industries thus far, despite being aware of the need to do so, and expressed their hope to continue this kind of activity.

4.6   Conclusion

This chapter argues that when using big data as a basis for people's decision-making and urban management, effective use of data can be achieved by accessing real life, that is, communicating beyond the data and the thoughts of the people behind the accumulation of the data. To realise this, we discuss the role and potential of living labs, where citizens from different walks of life can share their knowledge and improve and solve the facts and issues that exist in their lives using data. By organising the history of the development of Living Labs, the types, scope, and methods of activities, and case studies, it became clear that Living Labs provide participants with steps of exploration that open up different perspectives to each other. Through these steps, the participants could open up and discover the fundamental problems they felt they needed to solve. Although it is often said that the purpose of Living Labs is to create new value and innovation, the process should be step by step. For example, if a forest with a reputation for producing delicious mushrooms and wood is ideal for shelter, an amateur can easily get lost in the wood. To enjoy the forest's bounty, one must start at the entrance and gradually work further. It is vital to have courage to turn back, as there may be dangers along the way. We need to understand the whole environment that shapes the forest—the topography, soil, and climate—to create a map and build a road with much information to get the bounty and return it safely to our families. Thus, we can enjoy bounty and grow new resources. If we think of the trees in the forest as data, the soil as our lives, and the climate as a society (or perhaps we could call it economy), we will have a picture. The Living Labs served as a trekking guide. To avoid getting lost in a forest of data (trees) with too much expectation of benefits, we need to look at the essence of what we do with many eyes and a wide perspective.

References

1. Official Journal of the European Union 2003: Directive 2003/98/ec of the European parliament and of the council of 17 November 2003 on the re-use of public sector information (2003). https://eurlex.europa.eu/LexUriServ/LexUriServ.do?uri=OJ:L:2003:345:0090:0096:en:PDF. Accessed 29 Jan 2022
2. (2017) Enoll: https://enoll.org/about-us/. Accessed 29 Jan 2022
3. Adahl, M.: Commercial consortia. In: Living Labs, pp. 385–390. Springer, Berlin (2017)
4. Battleday, R.M., Peterson, J.C., Griffiths, T.L.: From convolutional neural networks to models of higher-level cognition (and back again). Ann. N. Y. Acad. Sci. **1505**(1), 55–78 (2021)
5. Chesbrough, H.: Open Services Innovation: Rethinking Your Business to Grow and Compete in a New Era. Wiley, New York (2011)
6. Dervin, B.: An overview of sense-making research: concepts, methods and results to date. The Annual Meeting of the International Communication Association (1983)
7. Hochreiter, S., Schmidhuber, J.: Long short-term memory. Neural Comput. **9**(8), 1735–1780 (1997)
8. Janssen, K., Dumortier, J.: Towards a European framework for the re-use of public sector information: a long and winding road. Int'l JL & Info Tech. **11**, 184–201 (2003)
9. Ohsawa, Y., McBurney, P.: Chance Discovery. Springer, Berlin (2003)
10. Robles, A.G., Hirvikoski, T., Schuurman, D., et al.: Introducing Enoll and Its Living Lab Community. European Commission, Brussels, Belgio (2015)
11. Smith, S.M., Ward, T.B., Finke, R.A.: The Creative Cognition Approach. MIT Press, Cambridge (1997)
12. Ståhlbröst, A.: A set of key principles to assess the impact of living labs. Int. J. Prod. Dev. **17**(1–2), 60–75 (2012)
13. Vaswani, A., Shazeer, N., Parmar, N., et al.: Attention is all you need. In: Advances in Neural Information Processing Systems 30 (2017)
14. Von Geibler, J., Erdmann, L., Liedtke, C., et al.: Exploring the potential of a German living lab research infrastructure for the development of low resource products and services. Resources **3**(3), 575–598 (2014)
15. Von Hippel, E.: Democratizing Innovation. MIT Press, Cambridge (2006)
16. Yu, H., Robinson, D.G.: The new ambiguity of open government. UCLA L. Rev. Discourse **59**, 178 (2011)

Part II
Explore, Collect, and Use Data

Chapter 5
Interpretability and Explainability in Machine Learning

Wai Kit Tsang and Dries F. Benoit

Abstract In machine learning, making a model interpretable for humans is becoming more relevant. Trust in and understanding of a model greatly increase its deployability. Interpretability and explainability are terms that refer to the understanding of a machine learning model. The relation between these two terms and the requirements for data mining tools are not always clearly defined. This chapter provides a framework for interpretability and provides a taxonomy of interpretability based on the literature. Properties of interpretability are related to the domain and to the methods involved. A distinction is made between inherently interpretable models and post-hoc interpretable models, which in the literature are also referred to as explainable models. This overview will argue that inherently interpretable models are more favorable for deployment than explainable models, which are not as reliable as inherently interpretable models.

5.1 Introduction

The impact of machine learning in our daily lives is undeniable. Decisions and recommendations are made by machine learning systems for millions of people around the world (e.g. Netflix or Amazon), while other decision processes take place more discretely in for example the automated steering of Boeing airplanes mid-flight. Failures of such systems can have drastic consequences, such as accidents caused by self-driving cars [23] or discrimination against racial groups in civil law, leading to unfair sentences [42]. For the past few years the European Union's General Data Protection Regulation severely restricts automated individual decision-making that "signifi-

W. K. Tsang · D. F. Benoit (✉)
Faculty of Economics and Business Administration, Ghent University, Tweekerkenstraat 2, 9000 Ghent, Belgium
e-mail: dries.benoit@ugent.be

W. K. Tsang
e-mail: waikit.tsang@ugent.be

D. F. Benoit
FlandersMake@UGent-CVAMO, Ghent, Belgium

89

cantly affect[s]" users [36]. This law legally mandates that users in machine learning systems have the "right of explanation," requiring a system to *explain* its decision-making process [13, 48]. This has led to the popular fallback of *interpretability*, which allows the verification of reasoning in machine learning systems [8].

5.2 What Is Interpretable Machine Learning?

Interpretability is not a trivial concept, since its meaning depends on the domain in which it is used and is closely related to how humans perceive and understand the world. The notion of interpretability in machine learning is elusive and cannot be *unambiguously* captured by a mathematical definition. An intuitive definition for interpretability is given by [30]: "Interpretability is the degree to which a human can understand the cause of a decision." This criterion already allows humans to gauge which model is more interpretable depending on how well the cause for its decision can be understood. To look for an interpretable model means seeking to understand its decisions.

Given that a model makes predictions with a high degree of accuracy, why is it not sufficient to simply trust the model and ignore the reasons for a certain decision? In many contexts there is a need for interpretability due to incompleteness of the problem formalization [8]. For certain problems or situations, it is not sufficient to obtain a prediction (the what); it is also necessary to understand how a certain prediction was reached (the why). In many tasks, reaching a correct solution is only part of the solution to your initial problem [32].

Many machine learning tasks require that a predictive task be not only highly accurate, but also safe in decision sciences such medical treatments in health sciences or the engineering of socio-technical systems, as for example in power plants [48]. In many sciences, machine learning is increasingly used for knowledge discovery, as for example in bio-informatics, where machine learning algorithms help in gaining insights into the biological processes of the human body [34]. Furthermore, the interpretability of a model can also aid in the detection of bias in machine learning models against, for example, discrimination against minorities [12]. In addition, during the development and debugging of a machine learning model, being able to interpret a model is valuable for diagnosing the models.

As a consequence of being able to interpret a model, certain traits can be checked more easily, such as fairness (ensuring that predictions remain unbiased) [16], privacy (protection of sensitive information) [9, 47], robustness (no overfitting on the noise in the input) [1], causality (checking for real relationships) [45], and trust (an interpretable model is easier to trust than a black box) [25]. In certain cases, however, interpretable models are not necessary, such as when the problem is well studied or when mistakes of the model do not have a severe impact [32].

5.3 A Taxonomy of Interpretability Models

Doshi-Velez and Kim [8] distinguishes common factors of interpretability on a domain level and on a method level using a data-driven approach. Let us consider a matrix where the rows are real-world tasks, columns are the specific methods, and each entry is the performance of the method on the end-task. An approach such as matrix factorization embeds both tasks and methods in low-dimensional spaces, as shown in Fig. 5.1, characterizing the similarities of tasks from a certain domain as well as for different methodologies. This approach finds the latent dimensions of interpretability for the domains and for the methods. The construction of such a matrix is expensive due to the evaluation of each cell in a real application by a domain expert. Hence, hypotheses are proposed about the latent dimensions.

5.3.1 The Task-Related Latent Dimensions of Interpretability

On a task-related level, each domain has their domain-specific requirements for interpretability. In a biological or physics context, researchers focus on scientific understanding (e.g., finding key features for galaxy formation), while in the banking sector decision makers look for a proper justification for a decision (e.g., explaining why a loan application was rejected). These examples illustrate the opposition of *global versus local* interpretability. Global interpretability aims to detect patterns on a general level [38, 46]. This level of interpretability denotes an understanding on a higher level taking into account its features, trained components, parameters, and general structure with a certain level of sparsity. It determines the relevant features for a model and is often is achieved by imposing a certain level of sparsity on the model. Any model that grows larger than a specific number of parameters or features quickly becomes too difficult for a human to mentally grasp [7, 29].

Local interpretability tries to justify a specific decision [2]. Local interpretability considers only one data point and attempts to understand why a certain prediction was made for a given input. By considering only one observation, the complexity

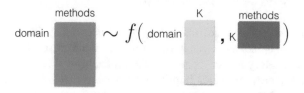

Fig. 5.1 This figure shows the decomposition using matrix factorization into embeddings for a matrix whose rows are real-world tasks, columns are the specific methods, and entries are the performance on the end-task. These embedding are lower-dimensional representations that contain factors that specify the interpretability factors for the domain level and for each method. Figure from [8]

of a model on a global level can be drastically diminished. Locally, the dependence among features in a model can behave entirely differently than on a global level [28, 35].

In certain tasks, the end user only has limited time to process and understand the explanation. In scientific applications, there might be many hours available to interpret the output, while in the operation of a chemical plant decisions have to be made more quickly. These *time constraints* vary from domain to domain. Furthermore, researchers and scientists are domain experts and have extensive user expertise, and thus explanations intended for them can be complex and sophisticated, while decision makers unfamiliar with the context, such as patients in healthcare, require simpler explanations. The *nature of user expertise* of the end-user in the task is strongly linked to the domain itself. This list of common factors characterizes a domain and is non-exhaustive.

5.3.2 The Method-Related Latent Dimensions of Interpretability

Similarly to the domain-specific explanations, disparate methods focus on diverse aspects of interpretability. This can be linked to the more precise notion of *model-based interpretability* and is defined by [33] as the *constraining of machine learning models* so that they readily provide useful information about the uncovered relationships. Reference [33], furthermore, divides the methods into two categories: the design of *inherently interpretable models* and the creation of *post-hoc interpretable models*, which take trained models as input and extract information about the relationships that the models have learned. Post-hoc interpretable methods are often applied when the underlying relationship is complex, and a black box model is used to obtain high accuracy. This chapter will primarily focus on the inherently interpretable models, and in later sections more explanations will be given as to why post-hoc interpretable methods are not always favorable.

5.3.2.1 Inherently Interpretable Models

Inherently interpretable models are designed by decreasing the complexity of a machine learning model and often have a simple and understandable structure [24]. These self-explanatory models incorporate interpretability directly into their structures and readily provide insight into the relationships they have learned. A list of (non-exhaustive) factors is provided here to address different requirements for interpretability based on [8, 32, 33].

- **Sparsity.** If the underlying relationship relies on a sparse set of signals, then it is possible to impose sparsity on a model by limiting the number of non-zero parameters. Given the form of the parameters and the magnitude of the set of non-zero parameters, the variables can be readily related to the output.

- **Simulatability.** A model is simulatable if a human being is capable of internally simulating and reasoning about a model's decision-making process. In other words, a human being can reason about how a trained model arrives at a certain output for a given input, which is possible if the feature space is small and the learned relationship simple. Decision trees are a good example of simulatability due to their hierarchical decision-making process [4].

- **Modularity.** A machine learning model is modular if a meaningful portion of its prediction-making process can be interpreted from the remainder of the model. Probabilistic models, for example, can impose conditional independence to facilitate reasoning about different parts of a model [22], and in generalized additive models, an additive relationship is imposed between variables in the model [17].

- **Domain-based feature engineering.** Creating informative features makes learning a model much simpler, allowing them to be more easier to interpret. In many domains, expert knowledge can be used to construct feature sets that are useful for predictive models.

- **Model-based feature engineering.** Many approaches automatically construct interpretable features, such as, for example, unsupervised learning and dimensionality reduction techniques. Principal component analysis (PCA) [20], independent component analysis (ICA) [3], and canonical correlation analysis (CCA) [18] are a few examples of model-based feature engineering.

- **Creating interaction between parts of the model.** Imposing monotonicity [14] and linear and non-linear interactions among parts of the model are a few examples to increase interpretability. Some functions are naturally more understandable for humans [43, 49].

- **Uncertainty and stochasticity.** The uncertainty measures can be improved so that the extent of stochasticity can be understood more easily. Confidence and prediction intervals are a few examples of how uncertainty can be quantified. Uncertainty measures that are provided with point forecasts, for example, can considerably increase the interpretability and deployability of a model.

5.3.2.2 Post-hoc Interpretable Models

While there has been an increase of studies on "explainable machine learning," where a post-hoc model is created to explain a black-box model, such explanations can be unreliable and very misleading. Making models and predictions explainable is highly domain-specific [11, 19], implying that there is no general domain-agnostic notion of interpretability, which should instead be constrained in model form to be useful to someone [14, 26]. There are many aspects of inherently interpretable models that could be inspected to make them more insightful for humans [32]. Although there has been a strong stream of literature aiming to work toward "explainable ML," where a second model is created *post-hoc* to explain the more complex black box model [39], there are some issues that need to be addressed.

5.4   Key Issues of Explainable ML

Instead of directly interpreting a trained function, explainable machine learning often builds a separate model to replicate the behavior of a black box posited to explain how a model works [37]. A black box model can be defined as (i) "a function that is too complicated for any human to comprehend," or (ii) "a function that is proprietary" [39]. These types of "explainable" models aim to explain how a model internally works, but do not necessarily reflect the world. The terminology of interpretable methods can lead to misleading conclusions, as these methods move away from the reality of problems in the world. Although in certain domains such a complex black box model could be a requirement, as in situations that require high-stakes decisions to be made, black box models are not necessary in areas such as criminal law [40] or risk assessment in the financial sector [6].

In addition, explainable machine learning methods do not always provide explanations that are faithful to the computation of a model. Any time a post-hoc interpretable model is created, it is never entirely faithful to the original computation. It is impossible to have an explanation that perfectly mirrors the original model, since this would mean that the explanation is exactly the same as the model and that the original model would thus be interpretable. This phenomenon leads to the real danger that an explanatory model for a black box can often inaccurately represent the original function, thus reducing trust in the explanation and by extension in the black box model [31]. Instead of creating explanations that fully reflect the original model, they show general patterns between the features and the predictions and should be considered more as summary statistics of predictions [39].

One such example where the explanation does not resemble the working of the actual model, would be the claim that the proprietary COMPAS (Correctional Offender Management Profiling for Alternative Sanctions) recidivism model was accused of being racially biased [10]. The linear explanation model was not capable of accurately explaining the decisions of COMPAS, since it reduced the reason for a decision sometimes to the race, while there might be correlation among other features [41]. Furthermore, in certain situations information outside the model should be combined with these black box models to obtain a calculation of risk. Certain external factors that cannot be incorporated in the model should be considered, but given the complexity of a black box, it is very difficult to manually adjust the prediction with additional information [5].

The emphasis is thus placed on inherently interpretable models, which provide their own explanations directly linked to what the model itself computes. The strong tendency to focus on the creation of black box models has resulted in high-stakes decision-making in domains like healthcare and criminal justice. By creating post-

Fig. 5.2 An application of explainable artificial intelligence to medical data. The model-agnostic explainable local interpretable method (LIME) is applied to detect lymph node metastases. Histological whole slide images were progressively divided into finer square grids. Explanation heat maps were generated for each of these grids, and a final heat map was computed on the basis previous heat maps. Image from [44]

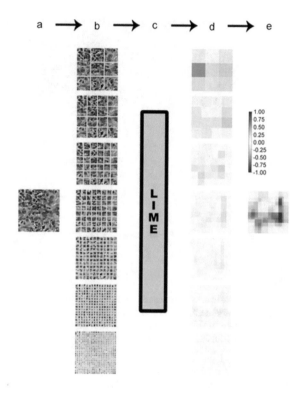

hoc interpretable methods for explaining these black box models, some of these problems were alleviated, but explaining black box models instead of creating models that are interpretable in the first place are likely to perpetuate bad practices and can potentially cause catastrophic harm to society [39].

As explained before, interpretability is relevant in many settings, especially in cases where high-stakes decision making is required. In medicine, for example, in the case of tumor detection in images, explaining what is visible on scans can be critical for medical diagnosis, as shown in Fig. 5.2. Similarly, in Massive Open Online Courses (MOOCs), student drop-out, i.e., when a student quits a course without completion, is notorious. Ideally, teachers and educational scientists will identify which students need extra help and aid them to prevent them from dropping out; see Fig. 5.3 for an example of drop-out analysis in lecture videos. Furthermore, when decisions are made for credit scoring, it is important for managers and decision makers to understand why certain predictions are made and how one should continue from these forecasts (See Fig. 5.4).

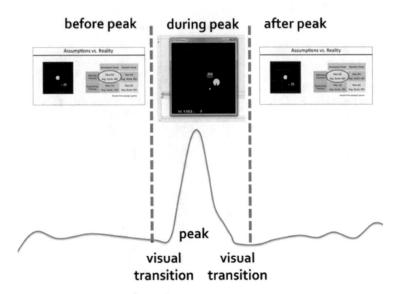

Fig. 5.3 When learners are watching the same online lecture videos, video-watching patterns can be analyzed to understand how students learn with videos. By using videos from Massive Open Online Courses (MOOCs) on edX, peaks in rewatching sessions and play events indicate points of interest and confusion in the lecture videos [21]

5.5 Conclusion

This chapter builds a framework for interpretable machine learning, mainly focusing on inherently interpretable machine learning instead of explainable machine learning. Interpretable methods ensure that predictions remain unbiased and make it easier for a human to trust a system that explains its decisions. This is especially crucial when there is something significant at stake such as the life of a human being. Interpretability in machine learning methods can be achieved by reducing the complexity of a model (intrinsic) or by applying methods to analyze the model after training (post-hoc). Model-specific tools are methods that only work for these specific models, while model-agnostic tools can be used on any machine learning model and are applied after the model has been trained (post-hoc). Individual predictions can be made more interpretable using so-called model-agnostic methods, and can be locally more dependent on a subset of features instead of a complex dependence on the feature space.

Irrespective of the interpretable methods used, it is important to evaluate and quantify the level of interpretability in machine learning. In order to make predictions interpretable, an explanation method draws a interpretable link between the feature values of an instance to its prediction. This chapter has provided an overview and taxonomy of interpretability based on the literature. Properties of interpretability are related to the domain and to the methods involved. This chapter has shown

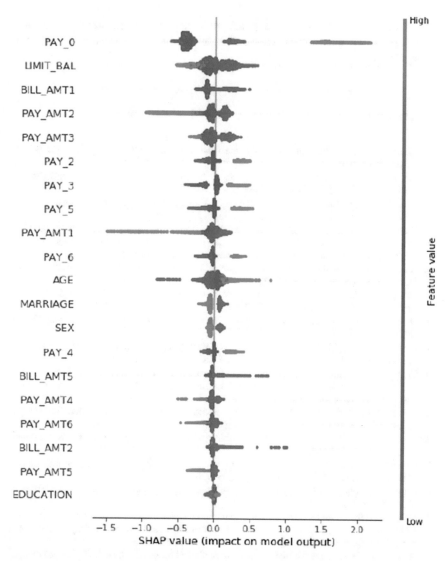

Fig. 5.4 Globally consistent Shapley summary plot for the UCL credit card dataset. SHAP (SHapley Additive exPlanations) is a game theoretic approach to explain the output for individual predictions [27]. The color for each observation is a representation of the value of the feature from low to high. As shown, the y-axis determines what the feature is and the x-axis is the Shapley value, i.e., the impact on the model output. A customer's most recent bill amount X_{BILL_AMT1} is one of the most important features globally, which aligns with reasonable expectations and basic domain knowledge. Image from [15]

that inherently interpretable models are more reliable and favorable for deployment than explainable models. Trust and understanding in a model contribute to its deployability.

References

1. Alvarez-Melis, D., Jaakkola, T.S.: On the robustness of interpretability methods (2018). arXiv:1806.08049
2. Baehrens, D., Schroeter, T., Harmeling, S., et al.: How to explain individual classification decisions. J. Mach. Learn. Res. **11**(Jun), 1803–1831 (2010)
3. Bell, A.J., Sejnowski, T.J.: An information-maximization approach to blind separation and blind deconvolution. Neural Comput. **7**(6), 1129–1159 (1995)
4. Breiman, L.: Classification and Regression Trees. Routledge, Abingdon-on-Thames (2017)
5. Brennan, T., Dieterich, W., Ehret, B.: Evaluating the predictive validity of the compas risk and needs assessment system. Crim. Justice Behav. **36**(1), 21–40 (2009)
6. Chen, C., Lin, K., Rudin, C., et al.: An interpretable model with globally consistent explanations for credit risk (2018). arXiv:1811.12615
7. Cowan, N.: The magical mystery four: how is working memory capacity limited, and why? Curr. Dir. Psychol. Sci. **19**(1), 51–57 (2010)
8. Doshi-Velez, F., Kim, B.: Towards a rigorous science of interpretable machine learning (2017). arXiv:1702.08608
9. Dwork, C., Hardt, M., Pitassi, T., et al.: Fairness through awareness. In: Proceedings of the 3rd Innovations in Theoretical Computer Science Conference, pp. 214–226 (2012)
10. Flores, A.W., Bechtel, K., Lowenkamp, C.T.: False positives, false negatives, and false analyses: a rejoinder to machine bias: there's software used across the country to predict future criminals. And it's biased against blacks. Fed. Probat. **80**, 38 (2016)
11. Freitas, A.A.: Comprehensible classification models: a position paper. ACM SIGKDD Explor. Newsl. **15**(1), 1–10 (2014)
12. Garcia, M.: Racist in the machine: the disturbing implications of algorithmic bias. World Policy J. **33**(4), 111–117 (2016)
13. Goodman, B., Flaxman, S.: European union regulations on algorithmic decision-making and a "right to explanation". AI Mag. **38**(3), 50–57 (2017)
14. Gupta, M., Cotter, A., Pfeifer, J., et al.: Monotonic calibrated interpolated look-up tables. J. Mach. Learn. Res. **17**(1), 3790–3836 (2016)
15. Hall, P.: On the art and science of machine learning explanations (2018). arXiv:1810.02909
16. Hardt, M., Price, E., Srebro, N.: Equality of opportunity in supervised learning. In: Advances in Neural Information Processing Systems, pp. 3315–3323 (2016)
17. Hastie, T.J.: Generalized additive models. In: Statistical Models in S, pp. 249–307. Routledge, Abingdon (2017)
18. Hotelling, H.: Relations between two sets of variates. In: Breakthroughs in Statistics, pp. 162–190. Springer, Berlin (1992)
19. Huysmans, J., Dejaeger, K., Mues, C., et al.: An empirical evaluation of the comprehensibility of decision table, tree and rule based predictive models. Decis. Support Syst. **51**(1), 141–154 (2011)
20. Jolliffe, I.T.: Principal component analysis and factor analysis. In: Principal Component Analysis, pp. 115–128. Springer, Berlin (1986)
21. Kim, J., Guo, P.J., Seaton, D.T., et al.: Understanding in-video dropouts and interaction peaks inonline lecture videos. In: Proceedings of the First ACM Conference on Learning@ Scale Conference, ACM, pp. 31–40 (2014)
22. Koller, D., Friedman, N.: Probabilistic Graphical Models: Principles and Techniques. MIT Press, Cambridge (2009)

23. Körber, M., Baseler, E., Bengler, K.: Introduction matters: manipulating trust in automation and reliance in automated driving. Appl. Ergon. **66**, 18–31 (2018)
24. Lakkaraju, H., Bach, S.H., Leskovec, J.: Interpretable decision sets: a joint framework for description and prediction. In: Proceedings of the 22nd ACM SIGKDD International Conference on Knowledge Discovery and Data Mining, pp. 1675–1684 (2016)
25. Lipton, Z.C.: The mythos of model interpretability (2016). http://arxiv.org/abs/1606.03490
26. Lou, Y., Caruana, R., Gehrke, J., et al.: Accurate intelligible models with pairwise interactions. In: Proceedings of the 19th ACM SIGKDD International Conference on Knowledge Discovery and Data Mining, ACM, pp. 623–631 (2013)
27. Lundberg, S.M., Lee, S.I.: A unified approach to interpreting model predictions. In: Advances in Neural Information Processing Systems, pp. 4765–4774 (2017)
28. Lundberg, S.M., Erion, G.G., Chen, H., et al.: Explainable AI for trees: from local explanations to global understanding (2019). http://arxiv.org/abs/1905.04610
29. Miller, G.A.: The magical number seven, plus or minus two: some limits on our capacity for processing information. Psychol. Rev. **63**(2), 81 (1956)
30. Miller, T.: Explanation in artificial intelligence: insights from the social sciences. Artif. Intell. **267**, 1–38 (2019)
31. Mittelstadt, B., Russell, C., Wachter, S.: Explaining explanations in ai. In: Proceedings of the Conference on Fairness, Accountability, and Transparency, pp. 279–288 (2019)
32. Molnar, C.: Interpretable Machine Learning (2019). https://christophm.github.io/interpretable-ml-book/
33. Murdoch, W.J., Singh, C., Kumbier, K., et al.: Interpretable machine learning: definitions, methods, and applications (2019). arXiv:1901.04592
34. Olson, R.S., La Cava, W., Mustahsan, Z., et al.: Data-driven advice for applying machine learning to bioinformatics problems (2017). arXiv:1708.05070
35. Plumb, G., Molitor, D., Talwalkar, A.S.: Model agnostic supervised local explanations. In: Bengio, S., Wallach, H., Larochelle, H., et al. (eds.), Advances in Neural Information Processing Systems 31, pp. 2515–2524. Curran Associates, Inc. (2018). http://papers.nips.cc/paper/7518-model-agnostic-supervised-local-explanations.pdf
36. Regulation, G.D.P.: Regulation (eu) 2016/679 of the european parliament and of the council of 27 April 2016 on the protection of natural persons with regard to the processing of personal data and on the free movement of such data, and repealing directive 95/46. Off. J. Eur. Union (OJ) **59**(1–88), 294 (2016)
37. Ribeiro, M.T., Singh, S., Guestrin, C.: Why should i trust you?: explaining the predictions of any classifier. In: Proceedings of the 22nd ACM SIGKDD International Conference on Knowledge Discovery and Data Mining, ACM, pp. 1135–1144 (2016)
38. Ribeiro, M.T., Singh, S., Guestrin, C.: Anchors: high-precision model-agnostic explanations. In: Thirty-Second AAAI Conference on Artificial Intelligence (2018)
39. Rudin, C.: Stop explaining black box machine learning models for high stakes decisions and use interpretable models instead. Nature Machine Intelligence **1**(5), 206–215 (2019)
40. Rudin, C., Passonneau, R.J., Radeva, A., et al.: A process for predicting manhole events in Manhattan. Mach. Learn. **80**(1), 1–31 (2010)
41. Rudin, C., Wang, C., Coker, B.: The age of secrecy and unfairness in recidivism prediction (2018). arXiv:1811.00731
42. Ruggieri, S., Pedreschi, D., Turini, F.: Data mining for discrimination discovery. ACM Trans. Knowl. Discov. Data (TKDD) **4**(2), 1–40 (2010)
43. Schulz, E., Tenenbaum, J.B., Duvenaud, D., et al.: Compositional inductive biases in function learning. Cogn. Psychol. **99**, 44–79 (2017)
44. Palatnik de Sousa, I., Maria Bernardes Rebuzzi Vellasco, M., Costa da Silva, E.: Local interpretable model-agnostic explanations for classification of lymph node metastases. Sensors **19**(13), 2969 (2019)
45. Spirtes, P.: Introduction to causal inference. J. Mach. Learn. Res. **11**(May), 1643–1662 (2010)
46. Tan, H.F., Song, K., Udell, M., et al.: Why should you trust my interpretation? Understanding uncertainty in LIME predictions (2019). http://arxiv.org/abs/1904.12991

47. Toubiana, V., Narayanan, A., Boneh, D., et al.: Adnostic: privacy preserving targeted advertising. In: Proceedings Network and Distributed System Symposium (2010)
48. Varshney, K.R., Alemzadeh, H.: On the safety of machine learning: cyber-physical systems, decision sciences, and data products. Big Data **5**(3), 246–255 (2017)
49. Wilson, A.G., Dann, C., Lucas, C., et al.: The human kernel. In: Advances in Neural Information Processing Systems, pp. 2854–2862 (2015)

Chapter 6
Interpretable GAM Models: Predicting Sepsis in ICU Patients

Wai Kit Tsang and Dries F. Benoit

Abstract Sepsis is one of the major causes of morbidity, mortality, and cost over-runs in critically ill patients. The survival of septic patients is greatly increased by early intervention with antibiotics. While machine learning models are helpful for health professionals in the diagnosis of patients, the lack of interpretability remains a barrier to the deployment of models in many healthcare applications. This study evaluates GAM models for sepsis prediction in terms of interpretability and predictive performance. Several variations of GAM models are included and offer a considerable improvement in predictive performance compared to logistic regression, while remaining interpretable. However, none of the GAM models achieved a predictive performance similar to that of a random forests classifier, which is highly effective but may not be sufficiently interpretable for end users. GAM models offer a balanced trade-off between explainability and predictive performance towards users.

6.1 Introduction

Early detection of sepsis is critical for improving patient survival rates in the intensive care unit (ICU). Sepsis, often referred to as "blood poisoning," arises when the body reacts to an infection in such a way that organs or other tissue are damaged [10]. Sepsis encompasses a wide variety of disorders, referred to as sepsis, severe sepsis, and septic shock, which can be caused by a diverse range of infections and dysregulated host responses. Septic shock was in its early phase described as a case of severe sepsis where hypo-tension occurs, and the concept of systemic inflammatory response syndrome (SIRS) was central to this description [28].

W. K. Tsang · D. F. Benoit (✉)
Faculty of Economics and Business Administration, Ghent University,
Tweekerkenstraat 2, 9000 Gent, Belgium
e-mail: dries.benoit@ugent.be

W. K. Tsang
e-mail: waikit.tsang@ugent.be

D. F. Benoit
FlandersMake@UGent-CVAMO, Ghent, Belgium

© Springer Nature Switzerland AG 2023
Y. Ohsawa (ed.), *Living Beyond Data*, Intelligent Systems Reference Library 230,
https://doi.org/10.1007/978-3-031-11593-6_6

Due to the need for more quantitative measures and more consistent reporting of sepsis occurrence, recent consensus definitions have been reached, resulting in Sepsis-3 [28]. To identify sepsis according to these new definitions, researchers are advised to use clinical criteria that identify life-threatening organ dysfunction [28]. The Sepsis-related Organ Failure Assessment (SOFA) score, a criterion developed by [31], intends to measure the degree of organ dysfunction by measuring low-complexity variables. To identify organ dysfunction in practice, it is defined as an increase in the SOFA score of at least 2 points.

Sepsis is associated with high mortality rates. In a global intensive care audit, a mortality rate of 35.3% was detected in patients with sepsis. Although the risk of death is higher for countries with a lower global national income, mortality rates are high in all countries [32]. Reference [10] measured a mortality rate of 17% for sepsis and 26% for severe sepsis in the period from 1979 to 2015 in high income countries. While results are hard to compare because of discrepancies between countries and the multitude of sepsis definitions, sepsis is estimated to be the global main cause of critical illness [28].

In addition, sepsis is also associated with a high societal cost. Reference [29] ranked sepsis as the most expensive medical condition in the United States, accounting alone for 5.2% of the national healthcare cost in the US in 2011. In the context of machine learning, not only the global cost is important. Since sepsis is very costly for the healthcare system, there is a need for cost-friendly and reliable screening and diagnostic methods [2]. One such tool that holds great potential are machine learning models that can accurately detect the early onset of sepsis [8]. While machine learning techniques have an immense potential for accurate diagnosis, their deployment by medical practitioners remains problematic. Due to the complexity of advanced machine learning models, there remains a large barrier to health care professionals bringing them into practice.

Whenever a medical decision is taken, it is important for the caregiver to motivate his decision. In situations where the health risks are high and the diagnosis is complicated, interpretable methods can bridge the gap between healthcare professionals and the complexity of machine learning tools. Interpretable machine learning is a branch of machine learning that specifically focuses on making predictions that can provide human-friendly explanations. The objective of this study is to analyze GAMs from the perspective of interpretable machine learning specifically applied to the prediction of sepsis in the ICU. GAMS have the benefit of imposing an additive relationship between variables in the model [13] so that each of the features can be interpreted separately in relation to the input.

The remainder of the chapter is organized as follows. First, the task is motivated by presenting disease diagnosis problems and the lack of explainability in machine learning models in recent studies. Interpretability in healthcare is unraveled and discussed and GAM models for interpretability are proposed and evaluated on a sepsis dataset extracted from the MIMIC-III database. The study is concluded with discussion and future perspectives.

6.2 Sepsis Prediction

6.2.1 Machine Learning in Healthcare

Since the early days of computing, decision support systems for clinicians have been developed. Decision support systems can assist clinicians in three domains: first, they enable attention focus by raising warnings if abnormalities appear. Second, it is also possible for decision support systems to deliver patient-specific predictions, leading to advice. Finally, scenario testing can be enabled by providing a guide to the most relevant historical data on the patient [18].

With advances in machine learning applications, these techniques have been implemented in the healthcare domain as decision support systems. The benefits of applying ML in healthcare are manifold. It enables the discovery of patterns across many observations that can be used to assist medical examiners in multiple ways. The model could automatically extract relevant information from the latest medical literature to support a clinician's decision. Furthermore, the system could help building an accurate diagnosis or advise on the therapy to be used. In addition, ML models could learn patterns in large amounts of patient data to discover predictions related to health outcomes [15].

Whereas machine learning and other AI technologies have been studied intensively for healthcare purposes, real-life application seems to be lagging [15]. The challenges for adopting machine learning technologies include regulatory, ethical, and legal issues. Legal issues include GDPR legislation, which obliges healthcare organizations to deliver explanations for predictions if the patients request them [1]. These challenges establish a need for interpretable machine learning models in healthcare applications. Interpretability of predictions is also required due to safety reasons: If data contain false patterns across patients that are taken into account by the model, model performance could be overestimated, leading to a higher proportion of false predictions that potentially are life-threatening [1].

6.2.2 GAMs for Sepsis Prediction

By far the most commonly used algorithms in ML applications for healthcare are neural networks and support vector machines [15], which are not considered interpretable [17]. With regard to sepsis prediction, [4] constructed an accurate machine learning classifier using gradient boosted trees. The model significantly outperformed traditional identifiers such as the MEWS, SOFA, and SIRS scores [5]. A clinical study was conducted that compared pre- and postimplementation mortality and length of stay. After implementing the model, sepsis-related mortality in the hospital dropped by 33.5%. Furthermore, the length of stays associated with sepsis dropped by 17%.

The need for interpretability was a recurrent theme both generally in healthcare-applied machine learning research and specifically in sepsis prediction. In order to make sepsis prediction more interpretable while maintaining good predictive performance, this study opts for the first approach of increasing the flexibility of an intelligible model.

Reference [6] proposed GAM models as intelligible models for healthcare. In their study, two healthcare-related problems were discussed: pneumonia prediction and prediction of re-admission into the hospital within 30 days. These intelligible model managed to uncover surprising patterns in the data that had previously remained undiscovered by complex models within this context [6]. Furthermore, GAMs constrain the variables in the model to have an additive relationship [13] so that each of the features can be interpreted separately with regard to the input.

Consequently, a gap in research can be identified. To the best of the author's knowledge, there have been no prior studies attempting to model sepsis classification with intelligible GAM models. In a study to benchmark deep learning models on the MIMIC database, [22] applied GAM models as a part of an ensemble that included logistic regression, neural networks, and random forests to predict sepsis. However, the focus of this study was on the benchmarking of the deep learning models. In addition, the ensembling rendered this model, called the Super Learner, non-intelligible [22].

6.2.3 Problem Definition

A gap in research is identified considering the use of interpretable GAM models to predict sepsis. The primary objective of this study is thus to construct an interpretable yet accurate GAM model for sepsis prediction. Furthermore, this study will focus on where this model is situated in the interpretability-performance trade-off. For the terminology on both interpretable machine learning and the difference between interpretability and explainability, the reader is referred to a previous chapter.

The goal of the model is to predict sepsis. Hence, it is a classification problem. The outcome variable is a binary variable, for which a value of 1 indicates the patient suffered from sepsis at that time and a value of 0 indicates no sepsis was present.

Section 6.3 will introduce the data used to build the GAM model. The MIMIC database offers a wide range of medical information, but a sepsis indicator is not readily available. Therefore, SOFA scores will need to be extracted from the database.

Methods applied in this dissertation will be discussed in Sect. 6.4. To position GAM models in the interpretability-performance trade-off, they will be compared to two benchmark models that hold a specific position on this trade-off. Details of GAM models, model building, and model evaluation are summarized. Finally, the results and discussion are addressed in Sects. 6.5 and 6.6.

6.3   Data

6.3.1   Origin

This study uses the MIMIC-III database [16]. MIMIC-III is a large database containing clinical data for almost 40,000 adult patients. The database includes a wide variety of information: medications, laboratory measurements, vital signs data, notes, diagnoses, procedures, and other information is stored in this database. The MIMIC database consists of the 26 tables, which can be categorized into four groups: definition and tracking of patient stays, critical care unit data, hospital record system data, and dictionaries.

The data were extracted from different tables within the database. The hospital database was used for information on the patient's stay in the hospital (tracking, procedures, prescriptions, personal details...). Two other tables were more specifically focused on the patient's stay in the intensive care unit (ICU). In one case data were also acquired externally: The date of death when occurring out of the hospital is derived from the social security registry. Data can also be categorized into static (e.g., admission time) or dynamic data (e.g., heart rate), the latter of which are periodically measured. Dynamic data have an associated itemid, whereas static data do not.

Since medical data are highly sensitive personal information, measures were taken to protect the privacy of the patients. All direct references to patient identification have been removed. As an additional measure, dates where shifted, resulting in dates of death up to four years in the future.

6.3.2   Identifying Sepsis

Identifying sepsis in the MIMIC database is not a trivial task. As a first approach, International statistical Classification of Diseases (ICD-9) codes were used to identify sepsis diagnoses. As suggested in the Sepsis-3 consensus definition [28], the code 995.92 signals sepsis as defined above. However, ICD-9 codes do not enable deducing the timing of sepsis onset. Since the aim of this analysis is to predict sepsis onset, the precise timing of this event is essential. Therefore, a more precise definition of sepsis was used. As explained in Sect. 6.1, sepsis can be detected by an increase in SOFA score of two points or more. The SOFA score is calculated over a period of time. The distribution of the SOFA scores of the observations included in the base table is shown in Fig. 6.1.

The SOFA score includes, among others, extreme values of some vital signs (e.g., heart rate) and laboratory results (e.g., platelet counts). As some of these variables have been reported to be good predictors of sepsis onset, they are included in the independent variable exploratory search. Of course, it is inevitable they will be correlated with the SOFA score if the value is included. Therefore, it will be important to strictly separate the independent from the dependent period.

Fig. 6.1 Histogram of
SOFA scores for
observations included in the
final base table

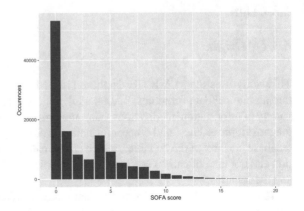

6.3.3 Data Extraction

6.3.3.1 Selecting Patients

This study only includes adult patients, i.e., patients who were at least 16 years old at the time of admission into the ICU. As defined in the Sepsis-3 convention, sepsis is detected if the SOFA score increases by two or more and there is a suspicion of infection. Patients are selected on the basis of a suspicion of infection during their stay in the ICU. Using the definition by [27], a suspicion of infection is characterized by both a blood culture draw and the prescription of an antibiotic. The first of these events will give the time of suspected infection. This information was retrieved from the prescriptions and microbiology events tables in the MIMIC database.

As shown in Fig. 6.2, this methodology resulted in a list of 32,300 relevant stays in the ICU. Studying only these ICU stays has the additional benefit of creating a dataset in which sepsis occurs in 39.27% of the cases. Moreover, it is reasonable to assume that in practice, a predictive tool would be used primarily when there is a suspicion of infection. Hence, this dataset could be considered representative.

6.3.3.2 Time Window

Following the methodology used by [8], data were collected for each ICU stay for which there was a suspicion of infection. In Fig. 6.3, the details of this strategy are visualized. A time interval is created around the first suspicion of infection occurring in the ICU stay. Throughout this interval, a SOFA score is then calculated at four separate points in time. The explanatory variables were collected during an independent period that does not overlap with the dependent period to ensure that these values are not represented in the dependent variable. For some variables, the value is assumed to be stable over this independent period (e.g., gender, age, and comorbidities). For bedside or laboratory measurements values are aggregated over the independent period, resulting in minimum, average, and maximum values.

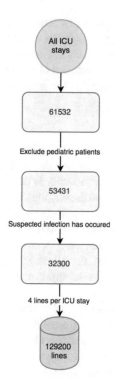

Fig. 6.2 The selection procedure for patients in the base table is illustrated here. The raw data pertain to 61,532 stays. After including only adult patients (at least 16 years old) and patients for whom there was a suspicion of infection, 32,300 stays remain for the final base table

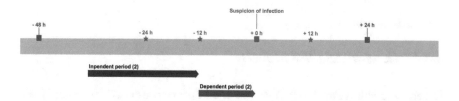

Fig. 6.3 Illustration of the timeline structure for a patient for whom sepsis occurs 24 h after the first suspicion of infection in the ICU stay. A patient is admitted to the ICU 48 h before the first suspicion and a SOFA score is then calculated at separate points in time. In this figure the SOFA score is calculated over a time interval with a 12-h dependent period and a 24-h independent period for the explanatory variables

The four observations were created by letting the dependent period start respectively 24, 12, and 0 h prior to first suspicion of infection, and 12 h after this event. Figure 6.3 shows how the observations are located in time. The independent and dependent period are shown for the second observation per ICU stay, for which the dependent period starts 12 h before the first suspicion of infection occurred.

6.3.3.3 Selecting Variables

Variables can be categorized into four groups: vital signs, laboratory results, demographics, and comorbidities. A comorbidity is defined as "a clinical condition that exists before a patient's admission to the hospital, is not related to the principal reason for the hospitalization, and is likely to be a significant factor influencing mortality and resource use in the hospital" [9]. In a first approach, all instances of these categories found in the MIMIC database were included. Consequently, the original base table consisted of over 90 variables. For all laboratory measurements, minimum and maximum values are included, and vital signs include minima, maxima, and average values.

In order to reduce dimensionality and computational load without losing too much information, some variables were excluded based on correlation statistics. For minima, averages, and maxima, correlations both among these variables and with the SOFA class target variable were compared. If there were positive correlations among the different instances for one item (e.g., heart rate), only the one correlated most highly with the SOFA class was included. Furthermore, variables that showed correlations of less than 0.01 with the target variable were excluded.

6.4 Methods

6.4.1 General Additive Models

Generalized additive models (GAMs) are a non-linear extension of generalized linear models (GLMs) [13] that demonstrate greater complexity than such GLMs as logistic regression. This results in a better predictive performance, while maintaining a high degree of interpretability, as explained in Sect. 6.2.

Several GAMs will be considered. First of all, standard GAM models are discussed. These standard GAM models should be understood as GAM models for which each variable is represented by exactly one shape function. The standard GAM models that will be discussed have varying shape functions.

Second, some extensions for the standard GAM are considered. Including pairwise interactions should increase the complexity of the predictor. GA^2M models include these terms. On the other hand, interpretability can be increased by selecting only those variables with the greatest predictive power. Several models have been suggested for this purpose, including generalized additive model selection (GAMSEL), sparse additive models (SpAMs), and sparse linear additive models (SpLAMs).

6.4.1.1   Standard GAMs

Standard generalized additive models can be represented by the set of functions shown in Eq. 6.2. Y_i is the outcome variable, x_j are the predictors, and g is the link function that transforms the sum of the shape function values into the predicted outcome variable. In the case of classification, g will be the sigmoid-shaped logit function.

$$E(Y_i) = \mu_i \tag{6.1}$$
$$g(\mu_i) = \beta_0 + f_1(x_1) + \cdots + f_n(x_n) \tag{6.2}$$

The shape functions f_j can be non-linear. In fact, a wide variety of shape functions can be used. Broadly, three types of shape functions can be distinguished: splines, trees, and ensembles of trees.

6.4.1.2   GAMs with Pairwise Interactions (GA^2M)

The performance gap between GAM models and full-complexity models such as random forests is due to the flexibility loss in GAM models, which only contain one-dimensional terms. Standard GAM models do not allow for the modelling of any interaction between variables. Therefore, [20] proposed a new class of GAM models that include interaction terms. GA^2M models are represented similarly to standard GAM models, but in addition to linear terms include one or more two-dimensional terms (Eq. 6.4).

$$E(Y_i) = \mu_i \tag{6.3}$$
$$g(\mu_i) = \beta_0 + \sum f_i(x_i) + \sum f_{ij}(x_i, x_j) \tag{6.4}$$

The main problem in building GA^2M models is the selection of relevant interaction variables. Even after selecting the most relevant one-dimensional features, there are still 1176 possible interaction features to consider in our model. To come up with the most relevant ones, [20] suggested a ranking algorithm, called FAST, that is based on the following procedure. First, a standard GAM model is built that includes all one-dimensional features. Next, for each candidate pair a model is trained to predict residuals of this model, thereby eliminating the main effect of both features. All pairs are then ranked according to performance in predicting the residuals. Because adding features one by one is computationally quite expensive for large datasets, [20] propose to select the top K interaction terms to include in the final GA^2M model. The parameter K can be fine-tuned.

6.4.1.3   General Additive Model Selection (GAMSEL)

Several models exist to limit the number of active predictors in a GAM model. We will apply GAMSEL, which is an extension to the SpAM model. Sparse additive models (SpAMs) shrink some variables to zero by applying a LASSO-related penalty term. The advantage of the SpAM model is it separates the penalty term and estimator, enabling SpAM for both spline-based and tree-based shape functions [23].

The generalized additive model selection (GAMSEL) model further extends this regularization by distinguishing predictors as either zero, linear, or non-linear. To do so, it builds on smoothing splines and thus is only applicable to spline-based GAM models [7]. Two penalty terms are included. Only the first penalty term will be discussed, since the second does not include any parameters that require tuning. The selection penalty includes the parameters α_j, which represent coefficients for linear terms. β_j are coefficients for non-linear terms (Eq. 6.5). Furthermore, the penalty contains two parameters. λ controls the extent to which parameters will be included in the model as non-zero. A high lambda will enforce greater regularization. The parameter γ specifically serves to distinguish linear from non-linear terms. For small values of γ, non-linear β_j coefficients will be over-penalized compared to linear coefficients, while for high values of γ, non-linear terms are favored over linear terms [7].

$$\lambda \sum_{j=1}^{p} (\gamma |\alpha_j| + (1 - \gamma)||\beta_j||_{D_j^*}) \tag{6.5}$$

6.4.2   Thresholds

Transforming probability scores to binary predictions requires a threshold. The possibilities for a threshold include:

- Standard 0.5 threshold
- Quantile of proportion of positives in training set
- Cost-minimizing threshold.

To evaluate the cost of using – or not using – a machine learning model, the cost of falsely positive and falsely negative classified observations should be estimated. The cost of a false positive or type II error was assumed to be the direct cost of treating a sepsis patient at the ICU. This involves costs for medication and other materials, but also personnel costs. Reference [3] estimated this cost to be 34,000 US dollars. The cost of a false negative or type I error is the additional cost of a late sepsis diagnosis compared to an early diagnosis. However, under a few assumptions an informed estimate can be made. Reference [21] compare costs for patients who are diagnosed upon arrival and those who are diagnosed later, during their stay at the ICU. For the latter, costs are 30,000 US dollars higher. They argue the difference in cost is due to higher mortality and more severe complications in case of a late diagnosis. Furthermore, [3] claim the direct cost of the treatment is only about 25% of the total cost of illness. Additional costs are primarily due to productivity losses resulting from a higher mortality. Therefore, it is assumed that the cost of failing to detect sepsis is four times the direct cost of 30,000 dollars. This resulted in following parameters.

$$Cost_{TypeI} = \$34,000 \tag{6.6}$$
$$Cost_{TypeII} = \$120,000 \tag{6.7}$$

Since the model aims to predict sepsis at the lowest possible cost, the cost-based strategy is used, which holds implications on the metrics used. Since a higher cost is associated with false negatives or type II errors, the cost-based approach will favor recall over precision. In all subsequent results, we will use the cost-based approach to report results.

6.5   Results

6.5.1   Model Evaluation

The performance of the GAM models is assessed using two benchmark models: logistic regression and random forest. To demonstrate the extent to which the models are interpretable, model outputs are given. Predictive performance is evaluated using AUC, precision, recall, and the f1-score. All models were trained on 80% of the dataset, while the remaining 20% is used as test data. Hyperparameters were tuned on the training set, using 5-fold cross validation. Classification results on the test set are shown in Tables 6.1 and 6.2.

Table 6.1 All models benched

Model	Type	AUC	Precision	Recall	F1-score
Logistic regression	L2 regularized	0.65	0.39	0.99	0.56
Logistic regression	Non-regularized	0.66	0.39	0.99	0.56
Random forest		0.90	0.77	0.86	0.81
Standard GAM	Spline-based	0.78	0.51	0.89	0.65
Standard GAM	Tree-based	0.79	0.51	0.89	0.65
GA^2M	Spline-based	0.78	0.51	0.90	0.65
GA^2M	Tree-based	0.79	0.52	0.89	0.66
GAMSEL		0.77	0.51	0.89	0.65

Table 6.2 Selected
interaction terms

Variable 1	Variable 2
Bicarbonatemin	DiasBPMean
BUNmax	RespRateMin
Chloridemax	Potassiummax
Chloridemax	TempCMax
Hemoglobinmin	MeanBPMean
INRmax	HeartRateMin
Lactatemax	DiasBPMean
Potassiummax	TempCMax
Sodiummin	DiasBPMean
WBCmax	SysBPMean

6.5.1.1 Interpretability

The logistic regression model delivers quite interpretable results. On the one hand, the L2 regularized estimated model, for which the variables were scaled in advance, allows the determination of the variables with the most predictive power. The coefficients of these variables will be higher, as illustrated by the 10 most important variables shown in Fig. 6.4. Since the variables were scaled before estimation, these can easily be compared.

To better understand the relationship between each predictor and the target variable, it is better to use non-scaled predictors. For now, two small examples will be discussed. The minimum number of platelet count, variable plateletmin in the regression, has a beta coefficient of -0.0016 in the non-restricted model. This implies if the minimum value platelet count goes up by 1 unit in the independent period for a patient, their odds of developing sepsis would be $e^{-0.0016} = 0.9984$ times as large. Similarly, the odds of developing sepsis when the maximum value of the anion gap increases by 1 would increase by more than five percent: $e^{0.0556} = 1.0572$.

When evaluating individual patients, finding the most important factors for their risk score is a matter of determining the contribution of each predictor in the final score, as demonstrated in Eq. 6.8. For this purpose, a scaled model should be used to account for the differences in variance among the different predictors.

$$\text{contribution}_i = \beta_i * \text{scaledpredictor}_i \tag{6.8}$$

Random Forest

For random forests, it is possible to retrieve the importance of each feature in the final model. However, these values of importance do not tell us anything about the functional form or even simply the direction of the relationship between the predictor and the outcome variable. To overcome this issue, partial dependence plots could provide an indication of the possible relationship between predictor and target. The

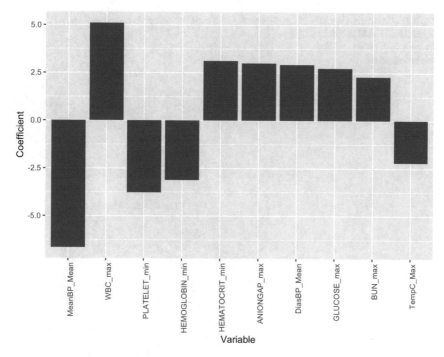

Fig. 6.4 Logistic regression delivers quite interpretable results. The coefficients of the 10 most important variables are shown here

partial dependence plot (PDP) shows the marginal effect of one or two features on the predicted outcome of a machine learning model [12]. While PDPs are intuitive and easily interpretable, they have the great disadvantage of assuming the independence of all features – that is, of assuming that the feature for which the PDP is calculated is not correlated with other features, which is fundamentally untrue for random forests. In addition, heterogeneous effects might be hidden, since PD plots only show the average marginal effects.

Standard GAM

To determine whether the GAM model is interpretable or not, interpretability will be discussed at two levels. First, the fitted shape functions will be discussed. These give a general overview of the relationship of each variable to the SOFA class target variable. The interpretability further increases if it is possible to rank predictors according to importance. On the level of individual predictions, a prediction can be considered interpretable if it is possible to identify which variables contributed most to the prediction. Preferably, it will allow us to deduce through which factor and in which direction a given predictor contributes to the final score for the patient.

On a general level, both spline- and tree-based GAM models achieve a good level of interpretability by visualizing the shape functions. In the case of spline-based

Fig. 6.5 A visualization of
the standard GAM predictors
for the spline-based and the
tree-based models

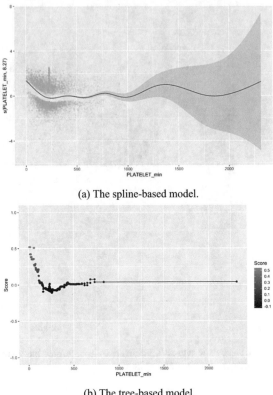

(a) The spline-based model.

(b) The tree-based model.

models, these are smooth functions. An example for the minimum platelet count is
given in Fig. 6.5a. The dark grey area shows the 95% confidence bounds for the fitted
spline. In areas with fewer observations, this interval grows substantially. However, in
the area between 0 and 500, enough observations are available to accurately estimate
the spline function, whose value is shown on the y-axis. This value represents the
contribution of the predictor to the final score before transformation by the logit
function. Apparently, levels of minimum platelet count in the lower ranges increase
the sepsis probability score.

The same predictor is visualized for the tree-based model in Fig. 6.5b. During the
implementation of this model, the predictors were discretized into a maximum of
256 bins before estimation. In Fig. 6.5b, the boundaries of the bins are shown as dots
connected by a line. The height of this line shows the value of the shape function
for predictor values within a given bin. Ranges located higher on the y-axis will thus
result in higher probability scores. A similar pattern to that in Fig. 6.5b is visible:
Although unusually high minimum platelet count values do not increase the sepsis
probability score, the contribution to the score rises as the minimum platelet count
drops for low values of plateletmin in the range of 0 to 200.

Fig. 6.6 The spline-based shape function for the maximum lactate values is shown. Around the mode of the variable (which indicates the missing values), a decrease in the spline function is visible

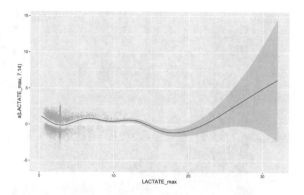

In Fig. 6.6, the spline-based shape function for the maximum lactate values is shown. Around the mode of the variable (which indicates missing values), a drop in the spline function is visible. This implies that not testing for maximum lactate values is related to a lower risk of sepsis development. This case immediately shows the benefits of the GAM model: It is more flexible than logistic regression and thus is able to detect such patterns in the data. On the other hand, it improves interpretability compared to random forests by detecting the direction of this relationship.

For individual predictions, the probability score is calculated simply by adding the function values $f(x_i)$ for the predictor variables x and transforming this sum to a probability, using the logit function. This enables us to rank the predictors x_i according to their contributions to the final score. If $f(x_i)$ is negative, the predictor x_i decreased the probability score for this observation. If it is positive, it contributed to the score.

The individual predictions also enable us to find the relative variable importance of the predictors. By aggregating the contributions of the predictors to each instance, an importance score can be obtained. For the tree-based GAM model, this resulted in the top 10 terms given in Fig. 6.7.

For the one-dimensional terms in the GA^2M model, interpretability is equivalent to a standard GAM model. Interaction terms can be visualized using heatmaps, of which an example is shown in Fig. 6.8. The interaction term ranked highest by the RANK algorithm is displayed: minimum that bicarbonate levels and mean diastolic blood pressure. The plot reveals high diastolic blood pressure will be a positive predictor for sepsis in case minimum bicarbonate levels are low, but will negatively contribute to the probability score when bicarbonate levels are high. Each of the interaction terms could be explained in a similar manner.

With the addition of interaction terms, there is no longer a single relationship between a predictor and a classification target. Although the interaction terms can be explained, this implies that the general intelligibility of the model will still slightly decrease.

For the GAMSEL model, because the number of terms is limited, not as many predictors should be discussed, leading to generally more interpretable results than

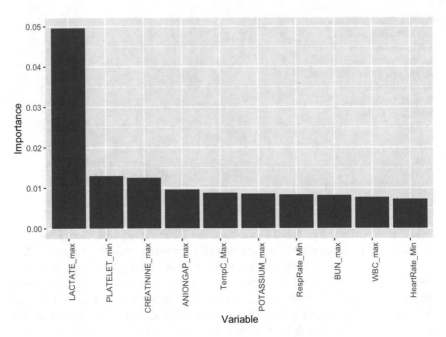

Fig. 6.7 The relative variable importance for the 10 most important predictors of the tree-based GAM model

Fig. 6.8 A heatmap visualizing the interaction term between minimum bicarbonate levels and mean diastolic blood pressure. The plot reveals high diastolic blood pressure as a positive predictor for sepsis in case minimum bicarbonate levels are low, but a negative one when bicarbonate levels are high

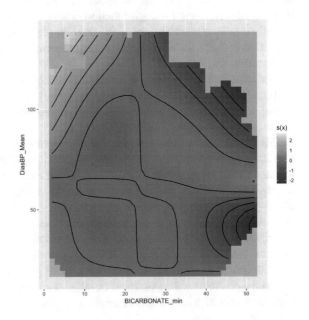

Table 6.3 Active binary, linear, and non-linear terms

Active binary	Linear	Non-linear	
M	Glucosemax	Aniongapmax	TempCmax
Surgery	PTmax	Choridemax	Creatininemax
Liver_disease	BUNmax	Lactatemax	Hemoglobinmax
Fluid_electrolyte	WBCmax	PTTmax	Potassiummax
Hypertension	Age	SysBPmean	HeartRatemin
Other_neurological		RespRatemin	MeanBPmean
Psychoses		Bicarbonatemin	Glucosemean
Chronic_pulmonary		Hematocritmin	Sodiummin
Paralysis		Plateletmin	DiasBPmin

standard GAM models. Linear terms can be more easily interpreted. The parameter β_i in the linear shape function (Eq. 6.9) should be interpreted as in logistic regression models: Ff x_i increases by one unit, the odds of sepsis become e_i^β larger (Table 6.3).

$$y_i = \alpha_i + \beta_i x_i \qquad (6.9)$$

6.5.2 Model Comparison

Models can be compared on three levels: the performance of the predictions, the interpretability of the estimation results, and finally the total cost of error.

The ROC curves give a good indication of the differences in performance of the models. The GAM models only slightly differ in performance, as shown in Fig. 6.9. GAMSEL has a slightly lower performance than the other models. Tree-based GAMs perform better than spline-based GAMs. Finally, GA^2M models perform a little better than their standard GAM counterparts.

As shown in Fig. 6.9, the ROC curves of the different GAM-based models do not differ greatly. The best-performing GAM model is a tree-based model with interactions. This model can be compared to the benchmark models (Fig. 6.10). The tree-based GA^2M model clearly performs better than a logistic regression, but cannot achieve the performance of the random forest model.

In the previous section, the difference in explainability between the benchmark and the different GAM models was demonstrated using some examples. Although logistic regression remains the simplest model, GAM models offer a serious improvement compared to random forests. Within the GAM modelling class, a GAMSEL model is slightly more interpretable than the others, due to the reduced number of graphs

Fig. 6.9 A comparison of
the ROC curves for GAM,
GA^2M, and GAMSEL.
GAMSEL has a slightly
lower performance than the
other models. Tree-based
GAMs perform better than
spline-based GAMs. Finally,
GA^2M models perform a
little better than their
standard GAM counterparts

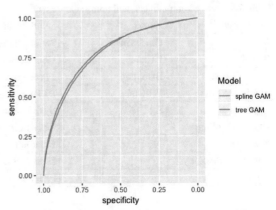

(a) The ROC curve for the standard GAM model.

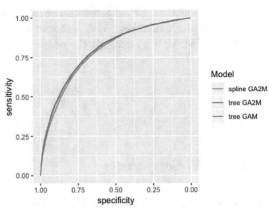

(b) The ROC curve for the GA^2M model.

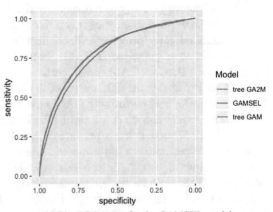

(c) The ROC curve for the GAMSEL model

Fig. 6.10 The ROC curves for the best GAM model compared with the benchmark models. The tree-based GA^2M model clearly performs better than a logistic regression, but cannot achieve the performance of the random forest model

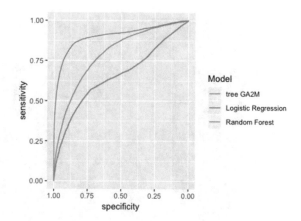

the reader has to examine. However, both standard GAM models and GA^2M models can be considered interpretable.

Finally, the cost of error is an important measure to decide which model should be used. Since this cost is closely related to the predictive performance, similar conclusions can be made. In Fig. 6.11, the total cost of error is shown for all models. Random forest stands out as the best model, whereas logistic regression has a considerably higher cost than all other models. In between these two, the various GAM models are comparable in cost of error. Nevertheless, tree-based GA^2M again stands out as the best GAM model in terms of modelling performance.

6.5.3   Focus: GA^2M Results

Formerly, predictive performance and examples of interpretability were emphasized without elaborating on the full modelling outcomes and implications for the sepsis classification problem. In this section, the focus will be on the output of the best-performing GAM model, which is the tree-based GA^2M.

The shape functions of the 10 most important parameters for the model are shown in Fig. 6.12, along with a short description of the pattern learned by the model for this feature. To interpret shape functions correctly, it is often necessary to understand the distribution of the input data and the values considered normal for the measure.

The most important interaction variable only ranks 26th in the aggregated contributions. This may explain why the GA^2M model only shows limited improvement in predictive performance compared to the standard GAM model.

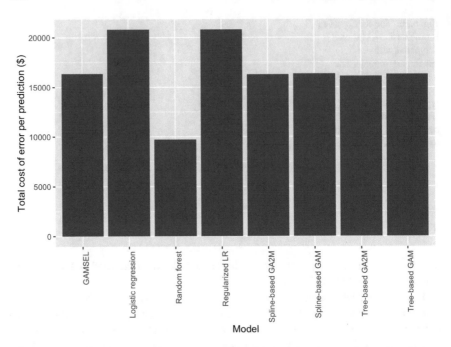

Fig. 6.11 The total cost of error is shown for all models. Random forest stands out as the best model, whereas logistic regression has a considerably higher cost than all other models. In between these two, the various GAM models are comparable in cost of error

6.6 Discussion

In this section, the most important results will be summarized. Results are two-fold: On the one hand, some takeaways for clinicians can be extracted from the GAM models. On the other hand, the focus is on the modelling results, more specifically the performance of GAM models with regard to the benchmark models will be compared to literature. In addition, this section addresses several limitations of this study and indications for future research.

6.6.1 Implications

Since features in the GAMs have an additive relationship, this allows the features to be investigated separately from each other, as opposed to complex black box models where the relation between input and output is not immediately comprehensible. In Sect. 6.5, the 10 most important predictors learned by the GA^2M model were shown and briefly discussed. From these visualizations, several conclusions can be made.

Fig. 6.12   The shape
functions of the 10 most
important parameters of the
tree-based GA^2M model are
shown here

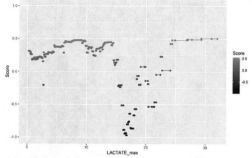

(a) $Lactate_{max}$. Values between 15 and 25 mmol/L show a considerable
drop in sepsis probability scores. However, data exploration shows that
most observations are located in the region under 10 mmol/L, where higher
values of maximum lactate are associated with higher sepsis probability
scores, with a drop around the mode.

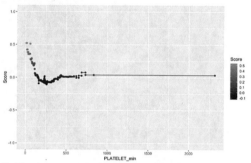

(b) $Platelet_{min}$. Low platelet count values of under 200 cells/mm3 correspond to increasing sepsis probability scores. A drop is visible around
250 cells/mm3. Platelet counts exceeding the normal range (over 400) do
not seem to contribute to sepsis probability.

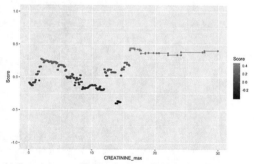

(c) $Creatinine_{max}$. High maximum creatinine levels will result in increased sepsis probability. In the lower, more common range of creatinine
levels between 0 and 10 mg/dL, lower levels of creatinine correspond to
lower sepsis probability. Around 5 mg/dL sepsis risk increases, and then
drops again as creatinine levels increase.

Fig. 6.12 (continued)

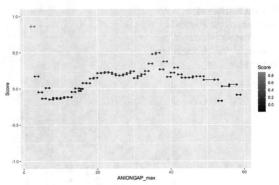

(d) *Aniongap$_{max}$*. In the prevalent area of maximum anion gap measures, higher values lead to higher sepsis probability scores.

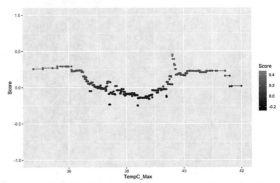

(e) *TempC$_{max}$*. Both lower and higher temperatures will result in a higher sepsis prediction score in the GA2M model.

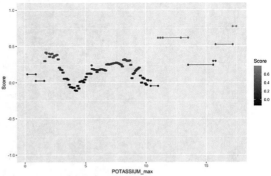

(f) *Potassium$_{max}$*. The model will give low sepsis probability scores to maximum potassium levels of approximately 4 mEq/L. Both divergent values below or above this level result in a higher sepsis probability score.

Fig. 6.12 (continued)

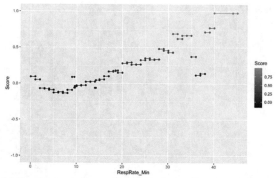

(g) $RespRate_{min}$. For respiration rates between 5 and 10 breaths per minute, GA2M estimates for sepsis probability are low. Lower values result in a slightly positive contribution to the sepsis score, whereas increased respiration rates heavily contribute to it.

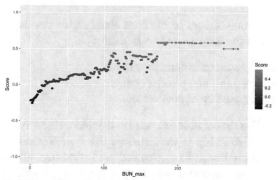

(h) BUN_{max}. Generally, higher blood urea nitrogen maximum values will lead to the model returning higher sepsis probability scores.

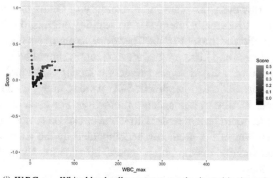

(i) WBC_{max}. White blood cell count was mostly situated in the region of 0 to 100 cells/μL in the data. There, the model will attach high sepsis probability scores to very low values of WBC counts. A small range of values results in negative contribution to the sepsis score, but as the WBC count increases, the sepsis score contribution grows.

Fig. 6.12 (continued)

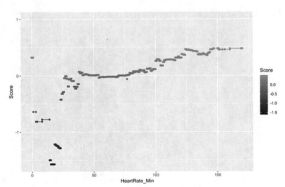

(j) $HeartRate_{min}$. Low minimum heart rates of under 40 beats/minute result in very low sepsis risk scores. Contrariwise, increased heart rates are associated with positive contributions to the score given by the model.

Clinicians should be cautious about sepsis development in ICU patients that have one of the following characteristics:

- Increased maximum lactate values.
- Low minimum platelet counts.
- Increased creatinine values.

This list could be extended until it includes every predictor, since for every variable the relationship is estimated by the GA^2M model. The benefit of GAM models precisely lies in its aggregated prediction over all variables. If a certain prediction is made, the score can be split out over all variables, ranked according to their contribution in this specific case. In this way, GAM models can offer an interpretable decision aid tool for clinicians.

However, the plain interpretability of GAM models should also be nuanced. In order to correctly interpret the modelling outputs, one has to take into account the distribution of the input data. Therefore, to report such a model to clinicians, descriptive statistics and histograms for each feature should be included. A concrete implementation of a GAM sepsis prediction model could include an application that shows the prediction for a specific ICU patient, along with the most important predictors contributing to this specific prediction. For each of the terms, additional information should be available within one click. This information should include the shape function of the predictor, accompanied by the score of the patient for this predictor, variable histogram(s), general information about the measure(s), and its normal values.

Regarding the position of GAM models on the interpretability-performance trade-off, modelling results were in line with the existing literature [19]: GAM models achieve better performance than logistic regression, but cannot reach the predictive performance of a full-complexity model such as random forests. On the other hand, GAM models offer an improvement in interpretability compared to random forests.

To successfully achieve an acceptable interpretability-performance trade-off, the predictive performance of the GAM models presented in this study does not suffice.

Whereas prior research [19] showed that tree-based GAM models perform better than the spline-based variant, this distinction is less clear in the sepsis prediction GAM models presented here. Although tree-based models performed better, the performance gains were only minor. Moreover, spline-based GAM models have the advantage of being slightly more interpretable [19]. Depending on whether interpretability or predictive performance is the primary objective of the model, both variants seem reasonable.

In order to improve predictive performance, researchers suggest adding interaction terms to the GAM models (GA^2M) [20]. In this study, adding 10 interaction terms resulted in a minor improvement of predictive performance. Furthermore, interpretability drops because the single relationship between predictor and outcome variable is lost.

One important addition to these implications is that a clear chasm exists between black box models and inherently interpretable models [24]. From the current results in the study, the random forest model outperforms the more interpretable GAM models. While predictive performance is highly relevant, this does not necessarily imply that a decision-maker blindly has to choose the most performant predictive model. It has been shown that complex unintelligible machine learning models currently used for high-stakes decision-making cause problems in, for example, criminal justice [33] and other domains such as healthcare [30]. Furthermore, explaining black box models instead of creating inherently interpretable models are likely to perpetuate bad practices since interpretability is a domain-specific notion [11, 14, 25]. Therefore, it is important to keep in mind that predictive performance should not necessarily be the final decision criterion when choosing a forecasting model, but its inherent properties for explainability should be taken into account as well.

6.6.2 Limitations

This study could be improved in several ways. A first limitation is situated in data extraction. To limit computation time, it was decided to extract only four lines per relevant ICU stay. However, more accurately detecting the time of sepsis onset would require more frequent observations, e.g., hourly. In addition, a gap could be set between the dependent period, in which SOFA is calculated, and the independent period that includes the predictors. By varying the length of this gap, it could be determined how long in advance the model is able to predict sepsis.

For the purpose of improving predictive performance, interaction terms were added to the model. Only the 10 most relevant interaction terms were added due to the high computation time for estimating these terms, specifically for spline-based methods [20]. Including a higher number of interaction features could improve the predictive performance of the GA^2M model. Furthermore, the relevant features were selected based on an execution of the RANK algorithm for a regression model (including plain SOFA score as outcome variable). To determine which features are most relevant, the RANK algorithm could be adapted to be compatible with classification modelling tasks.

Improvements could also be made in the evaluation of the models. In this study, the average cost of error per prediction was taken as a metric to discriminate between models. However, correct predictions can also result in further benefits than in a situation in which no model is used. This is true both for correctly classified positive predictions, in which early detection can avoid hospital costs, and for correctly classified negative predictions, for which additional observations, increased length of stay, and higher medication costs can be avoided. Taking into account the benefits of earlier detection could offer a more holistic evaluation measure, which is also more attractive to healthcare organizations interested in the deployment of such a model.

6.6.3   Future Research

Suggestions for future research include improvements related to the limitations mentioned above. More specifically, the usefulness of GAM models in sepsis prediction focusing on onset timing could be a valuable contribution, as this study did not include enough observations to achieve an accurate estimate of sepsis onset timing.

Furthermore, while complex models can achieve high predictive performance, models that are accurate and interpretable exist in many domains [24]. Reference [26] argues that interpretable models exist across multiple domains but may be hard to find through optimization. While interpretable machine learning looks to the extraction of explanations from existing high-performing prediction models, efforts should be continued to investigate inherently interpretable models. Deploying explainable black-box models in high-stakes decision-making can be harmful for policy-makers in the long run and lead to deterioration of safety and trust in machine learning models.

6.7   Conclusion

The objective of this study is two-fold: to build an interpretable GAM model for sepsis prediction and to situate it in the interpretability-performance trade-off. The research question thus includes whether a GAM model could approximate the predictive performance of a full-complexity model while producing interpretable results.

In order to evaluate both the interpretability and the predictive performance of GAM models, two benchmark models were selected that represent the extremes of the trade-off: logistic regression and random forest, respectively, for high interpretability and predictive performance. The performance of these models was compared to various GAMs: spline- and tree-based GAMs, GAMs with interaction pairs (GA^2M), and regularized GAMs (GAMSEL). All models were evaluated based on the average cost of error per prediction. As reported by other studies, the tree-based GA^2M outperformed the other GAM models. However, the performance differences were rather small. None of the GAM models achieved predictive performance levels close to those of random forests.

In terms of interpretability, GAM models perform well. The features in GAMs can be interpreted separately from each other due to the additivity in the feature space. However, for a fully comprehensive explanation of the modelling results, information on the input data is required. We thus conclude that the predictions of GAM models can be explained to clinicians on the condition enough information about the individual features and their distribution is provided. In practice, this information should be gathered in an application in order to present interpretable predictions to clinicians in the ICU.

In the context of such application, GAMs are well suited to assist clinicians in sepsis treatments by providing interpretable predictions. However, the predictive performance gap with full-complexity models has to disappear in order for GAMs to become practically relevant. Adding more two-dimensional interaction terms can potentially help close this gap. However, it is also possible that two-dimensional interactions are not sufficient to model sepsis development. Therefore, future research efforts can be directed either to improving GAM predictive performance, or to enhancing the interpretability of more complex models. This challenge of interpretability-performance trade-off is a universal issue in machine learning, both in healthcare and in other domains.

Acknowledgements The authors would like to thank Birgit Mullie whose masterthesis served as basis for this chapter.

References

1. Ahmad, M.A., Eckert, C., Teredesai, A.: Interpretable machine learning in healthcare. In: Proceedings of the 2018 ACM International Conference on Bioinformatics, Computational Biology, and Health Informatics. ACM, pp. 559–560 (2018)
2. Angus, D.C., Linde-Zwirble, W.T., Lidicker, J., et al.: Epidemiology of severe sepsis in the united states: analysis of incidence, outcome, and associated costs of care. Crit. Care Med. **29**(7), 1303–1310 (2001)
3. Burchardi, H., Schneider, H.: Economic aspects of severe sepsis. Pharmacoeconomics **22**(12), 793–813 (2004)
4. Burdick, H., Pino, E., Gabel-Comeau, D., et al.: Evaluating a sepsis prediction machine learning algorithm in the emergency department and intensive care unit: a before and after comparative study. bioRxiv, p. 224014 (2018)
5. Burdick, H., Pino, E., Gabel-Comeau, D., et al.: Effect of a sepsis prediction algorithm on patient mortality, length of stay, and readmission. bioRxiv, p. 457465 (2018)
6. Caruana, R., Lou, Y., Gehrke, J., et al.: Intelligible models for healthcare: predicting pneumonia risk and hospital 30-day readmission. In: Proceedings of the 21th ACM SIGKDD International Conference on Knowledge Discovery and Data Mining. ACM, pp. 1721–1730 (2015)
7. Chouldechova, A., Hastie, T.: Generalized additive model selection (2015). arXiv:1506.03850

8. Desautels, T., Calvert, J., Hoffman, J., et al.: Prediction of sepsis in the intensive care unit with minimal electronic health record data: a machine learning approach. JMIR Med. Inform. **4**(3), e28 (2016)
9. Elixhauser, A., Steiner, C., Harris, D., et al.: Measures for use with administrative data comorbidity. Med. Care **36**(8), 27 (1998)
10. Fleischmann, C., Scherag, A., Adhikari, N.K., et al.: Assessment of global incidence and mortality of hospital-treated sepsis. current estimates and limitations. Am. J. Respir. Crit. Care Med. **193**(3), 259–272 (2016)
11. Freitas, A.A.: Comprehensible classification models: a position paper. ACM SIGKDD Explor. Newsl. **15**(1), 1–10 (2014)
12. Friedman, J.H.: Greedy function approximation: a gradient boosting machine. Ann. Stat. pp. 1189–1232 (2001)
13. Hastie, T.J.: Generalized additive models. In: Statistical models in S. Routledge, p. 249–307 (2017)
14. Huysmans, J., Dejaeger, K., Mues, C., et al.: An empirical evaluation of the comprehensibility of decision table, tree and rule based predictive models. Decis. Support Syst. **51**(1), 141–154 (2011)
15. Jiang, F., Jiang, Y., Zhi, H., et al.: Artificial intelligence in healthcare: past, present and future. Stroke Vasc. Neurol. **2**(4), 230–243 (2017)
16. Johnson, A.E., Pollard, T.J., Shen, L., et al.: Mimic-iii, a freely accessible critical care database. Sci. Data **3**(160), 035 (2016)
17. Kotsiantis, S.B., Zaharakis, I., Pintelas, P.: Supervised machine learning: a review of classification techniques. Emerg. Artif. Intell. Appl. Comput. Eng. **160**, 3–24 (2007)
18. Lisboa, P.J.: A review of evidence of health benefit from artificial neural networks in medical intervention. Neural Netw. **15**(1), 11–39 (2002)
19. Lou, Y., Caruana, R., Gehrke, J.: Intelligible models for classification and regression. In: Proceedings of the 18th ACM SIGKDD International Conference on Knowledge Discovery and Data Mining. ACM, pp. 150–158 (2012)
20. Lou, Y., Caruana, R., Gehrke, J., et al.: Accurate intelligible models with pairwise interactions. In: Proceedings of the 19th ACM SIGKDD International Conference on Knowledge Discovery and Data Mining. ACM, pp. 623–631 (2013)
21. Paoli, C.J., Reynolds, M.A., Sinha, M., et al.: Epidemiology and costs of sepsis in the united states-an analysis based on timing of diagnosis and severity level. Crit. Care Med. **46**(12), 1889 (2018)
22. Purushotham, S., Meng, C., Che, Z., et al.: Benchmark of deep learning models on large healthcare mimic datasets (2017). arXiv:1710.08531
23. Ravikumar, P., Lafferty, J., Liu, H., et al.: Sparse additive models. J. R. Stat. Soc. Ser. B (Statistical Methodology) **71**(5), 1009–1030 (2009)
24. Rudin, C.: Stop explaining black box machine learning models for high stakes decisions and use interpretable models instead. Nat. Mach. Intell. **1**(5), 206–215 (2019)
25. Rüping, S.: Learning interpretable models. Ph.D. thesis, TU Dortmund University (2006)
26. Semenova, L., Rudin, C.: A study in rashomon curves and volumes: a new perspective on generalization and model simplicity in machine learning (2019). arXiv:1908.01755
27. Seymour, C.W., Liu, V.X., Iwashyna, T.J., et al.: Assessment of clinical criteria for sepsis: for the third international consensus definitions for sepsis and septic shock (sepsis-3). Jama **315**(8), 762–774 (2016)
28. Singer, M., Deutschman, C.S., Seymour, C.W., et al.: The third international consensus definitions for sepsis and septic shock (sepsis-3). Jama **315**(8), 801–810 (2016)
29. Torio, C., Andrews, R.: National inpatient hospital costs: the most expensive conditions by payer, 2011: statistical brief# 160 (2006)
30. Varshney, K.R., Alemzadeh, H.: On the safety of machine learning: cyber-physical systems, decision sciences, and data products. Big Data **5**(3), 246–255 (2017)
31. Vincent, J.L., Moreno, R., Takala, J., et al.: The sofa (sepsis-related organ failure assessment) score to describe organ dysfunction/failure. Intensiv. Care Med. **22**(7), 707–710 (1996)

32. Vincent, J.L., Marshall, J.C., Namendys-Silva, S.A., et al.: Assessment of the worldwide burden of critical illness: the intensive care over nations (icon) audit. Lancet Respir. Med. 2(5), 380–386 (2014)
33. Wexler, R.: When a computer program keeps you in jail: How computers are harming criminal justice. New York Times 13 (2017)

Chapter 7
Undesigned Data in Discovery Processes and Design of Their Interpretation

Akihiro Yamamoto and Sae Kondo

Abstract In this study, we stress the importance of treating undesigned data to discover knowledge. The category of undesigned data is clarified by revisiting knowledge discovery in databases (KDD), process models, and hypotheses in each of the models. We propose the treatment of undesigned data by referring to mathematical logic. We also show that the process model of KDD can be refined by referring to the process models of fields other than KDD.

Keywords Agile model · KDD process · Hill-climbing model · Hypothesis generation · Hypothesis validation · Interpretation of data · PPDAC cycle · Process model · Undesgined data · Software engineering

7.1 Introduction

Various methods for analyzing data have been developed and improved in the field of machine learning. When we apply them to knowledge discovery in databases (KDD), we often notice gaps between analyzing data and extracting knowledge from data. For example, neural networks, which are now attracting considerable attention, accept only numerical vectors as data, and the results of their learning using such data are parameters in the networks. Trained neural networks work very well when applied to newly obtained data. However, it still quite difficult to answer the question "What is learned by training a neural network?" or "What is the knowledge that a trained neural network has?" because the parameters are numerical; therefore, they must be interpreted when constructing knowledge. Recently, it has become increasingly

A. Yamamoto (✉)
Graduate School of Informatics and Center for Innovative Research and Education in Data Science, Kyoto University, Kyoto, Japan
e-mail: akihiro@i.kyoto-u.ac.jp
URL: http://www.iip.i.kyoto-u.ac.jp/akihro/index.html

S. Kondo
Department of Architecture, Graduate School of Engineering, Mie University, Mie, Japan
e-mail: skondo@arch.mie-u.ac.jp

© Springer Nature Switzerland AG 2023
Y. Ohsawa (ed.), *Living Beyond Data*, Intelligent Systems Reference Library 230,
https://doi.org/10.1007/978-3-031-11593-6_7

difficult to interpret the obtained parameters because the number of parameters in a neural network has increased. In addition, when non-numerical data are applied to neural networks, they must be translated into numerical vectors. This means that we need to interpret translation when extracting knowledge. In this article, we present a viewpoint on the gap between data analysis and knowledge extraction. The point is the process model of the KDD. From this viewpoint, we distinguish two types of data, designed and undersigned, and then discuss the interpretation of undesigned data by referring to the interpretation of sentences in mathematical logic.

In the next section, we introduce two projects that one of the authors proceeded with previously as concrete examples of providing interpretation to undesigned data. This would help readers understand the discussion. In Sect. 7.3, we first point out that at least two types of processes have been formalized and proposed for KDD. The KDD process is formed with smaller steps of subprocesses. We claim that the two types of data should be distinguished depending on the role of the hypotheses in each type of KDD process. The first type of KDD process is presented in the rise of KDD research, which we call the hill-climbing model in this article. The second is the PPDCA cycle, which is well referred to in the education of data science (see textbooks on data science, e.g., [14]) and models traditional research processes in natural science, social science, physiology, and so on. In the PPDCA cycle, we first define the problem that we solve and generate the hypotheses. Then, we designed the data for analysis, collected them, and eventually validated the hypotheses. In the hill-climbing model, we obtain the data first and then generate hypotheses; therefore, the data are not always designed for the analysis. From the comparison of the two types of processes, we find that our problems are how to improve the processes and how to manipulate the undersigned data. Undesigned data treated in the hill-climbing model are those obtained by sensors without specific purposes or used for analysis, which is not intended when they are generated. In Sect. 7.4, we propose to regard them as sentences when we adopt the hill-climbing model for KDD in the sense of mathematical logic [1, 8] and formal language theory [5]. In particular, mathematical logical sentences are clearly separated from their interpretations. In the PPDCA cycle, data are designed for a specific analysis; therefore, they have their own interpretation, whereas in the hill-climbing model, the interpretation of data is provided in the analysis. In Sect. 7.5, we propose that the hill-climbing model can be improved by comparing it with processes in other areas. In various fields, we can find an analysis of work for certain purposes by formalizing it as process models. The name "hill-climbing" is from its similarity to the famous "waterfall" model in software engineering [11, 13]. In the development of software engineering, the waterfall model is used to manage resources, including humans, and to clarify the test process of the produced software. Process models are formalized in other productions, not only related to computers, but also to large or complex products such as buildings. By comparing KDD processes with those in other fields, we noticed that some types of processes are not clearly formalized or proposed. In the last section, we present our discussion and remarks.

7.2 Two Projects as Concrete Examples for Our Discussion

We introduce two previous projects which we refer to in the discussion later.

7.2.1 Discovering Interesting Travel Routes from Trajectory Data

The first project was to discover interesting travel routes from trajectory data in Japan, which records the trajectories of some tourists or vehicles [9]. In this study, we refer to this project as *trajectory data mining*. The trajectory data were constructed with the mobile phones carried by them and GPS technology. They are stored in the form of a simple relational table whose schema consists of attributes "Terminal ID," "latitude," "longitude," "date," and so on as Table 7.1. The schema would have been designed from the viewpoint of physics, where every trajectory is a sequence of pairs of times and places. This means that the knowledge discovered from the data should be in the terminology of physics, but we cannot say that such knowledge is useful, for example, for tour agencies or railway/bus companies. In their business, for example, designing tour courses for travelers, they would use not only physical knowledge, such as the velocity of vehicles, buses, or trains, but also geographic and historical knowledge. This means that we have to discover knowledge represented in the terminology, not physics, but geography. To solve this problem, it was proposed in [9] to transform the original trajectory data into another form using geographic background knowledge. More precisely, each pair of latitude and longitude values in a trajectory is transformed into a text datum that presents its address in the Japanese addressing system. The system is useful in representing knowledge discovered by the method because most Japanese addresses are represented with a hierarchical structure consisting of several levels: the first level is for prefectures, the second is for cities, and the third is for districts in cities. By fixing the level as a parameter before applying our discovery algorithm, knowledge of the geographically appropriate resolution can be obtained. An example of frequent sequences discovered is:

Table 7.1 A small fragment of trajectory data

TerminalID	Latitude	Longitude	Date
a9563341-***	35.84815	139.7927	2018/12/31 22:42
a9563341-***	35.84767	139.7913	2019/1/1 0:24
e16b1686-***	35.67364	139.7699	2018/12/31 23:26
e16b1686-***	35.6736	139.7699	2018/12/31 23:41
e16b1686-***	35.67362	139.7699	2018/12/31 23:55

(*Nishi-Shinjuku, Shinjuku, Tokyo*)

⟶ (*Tsuru, Tsuru-City, Yamanashi*)
⟶ (*Fujimi, Fujiyoshida city, Yamanashi,Japan*)
⟶ (*Yosehoncho, Midori Ward, Sagami City, Kanagawa*)
⟶ (*Jingumae, Shibuya, Tokyo*)

where each triplet consists of the names of the districts, cities, and prefectures. By referring to some geometrical information, we can see that there is a large bus terminal in the Nishi-Shinjuku district in Tokyo, and Fujimi is a good place for sightseers to watch Mt. Fuji. The fact that the sequence often appears in trajectory data implies that many tourists move along the sequence. We could conjecture that tourists do not take trains by consulting Japanese railway maps and could also find that bus tours are operated along it.

7.2.2 Collecting and Unifying Syllabuses of Various Universities

The second project is *syllabus data integration*, motivated by the "AI Strategy 2019",[1] which was established by the Japanese government. In this strategy, the government declares that, by 2025, they will make and implement plans so that all students of all universities in Japan can learn the literacy of data science. Achieving this plan should have started by investigating the current situation of data science education in universities. The method collects as many syllabi as possible from the homepages of various universities and integrates information in a uniform format so that we can answer questions about the situation (Fig. 7.1), for example, "Which key words are often correlated to courses named in Computer Science?" and "How many faculties have any courses on Mathematical Statistics?" The recipe for knowledge discovery is as follows: collecting syllabuses of courses opened on the homepages of universities and colleges, transforming each of them into a unique format, and storing it in a database management system. The problem that we had to solve is that the format of homepages for syllabi is not standardized. Their views on browsers vary from university to university. Even if the two views appear similar, their source codes in the HTML would be different. The solution in [6] applies a neural network method to HTML source codes and extracts some attributes common to all syllabi. Because the minimum set of attributes that each syllabus must contain is defined by the ministry of the government, extracting a subset of the minimum set needed for the investigation is sufficient. In addition, every university has its own format of the HTML source code of syllabus pages. Several pages were chosen for the training data of the machine learning, and the trained neural network was applied to other pages. The learning

---

[1] https://www.kantei.go.jp/jp/singi/ai_senryaku/pdf/aistratagy2019.pdf (in Japanese).

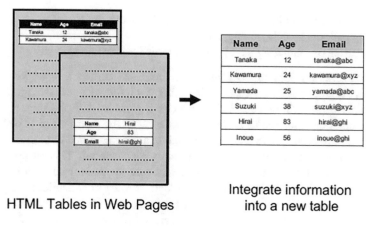

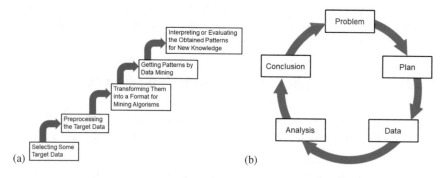

HTML Tables in Web Pages   **Integrate information into a new table**

Fig. 7.1 Integration of syllabuses from various universities

method is a modification of the modern method for natural language processing so that it can treat HTML codes, which are a mixture of artificial tags and sentences in natural languages.

7.3   Models of KDD Process

7.3.1   Two Models of KDD Process

The KDD process was originally modelled in the 1990s. The process in [2] is illustrated in Fig. 7.2a and consists of five steps:

– selecting some target data,
– prepressing the target data,

(a)   (b)

Fig. 7.2 Two KDD Process models, hill-climbing (**a**) and the PPDAC cycle (**b**)

- transforming them into a format for mining algorisms,
- getting patterns by data mining, and
- interpreting or evaluating the obtained patterns for new knowledge,

where the original data are located in the lower-left corner, and the knowledge obtained eventually is in the upper-right corner. By referring to the *waterfall* model in software engineering (see Sect. 7.5) we call this model *hill-climbing*. Note that in the generation when the model was proposed, both the Internet and WWW were still under development and not popular, and data were assumed to be stored in the DBMS in some mainframe computers owned by each institute. This causes the process to start by selecting data, which may be difficult to obtain from other institutes. This situation might still be present in the data that are confidential to any other organization. Recently, in the field of data science education, a cyclic model has been admitted to teach students what data science is and what data is. The model is illustrated as a cyclic connection of five steps: problem, plan, data, analysis, and conclusion. Therefore, the cycle is called the PPDAC cycle. In the following list, the tasks that should be performed at each step are presented[2]:

- Problem step: We must understand and define the problem or question, and determine a potential strategy for answering it.
- Plan step: We must determine how to measure and develop our study design.
- Data step: We collected data, cleaned data, and addressed any data issues.
- Analysis step: We produced tables, graphs, and summary statistics, and analyzed data specific to the question and study design.
- Conclusion step: We interpret the results, communicate the findings, and generate ideas for future analysis.

It should be noted that this can be regarded as an abstract model of traditional academic research areas using data, such as physics, economics, and psychology; therefore, we compared it with the hill-climbing model as a formulation of research processes. The most important difference between the two is in the problem. The PPDCA cycle is initialized in the problem step. This means that the cyclic process cannot transit to any other step until the problem to be solved is not well-defined. On the other hand, the hill-climbing model does not have a problem step; therefore, the process can start before the problem to be solved is defined. This difference arises from the two types of research activities explained in the following subsection.

7.3.2 Hypothesis Validation or Hypothesis Generation

We found at least two types of research processes in academic research. The first is *hypothesis validation*. This approach has been adopted in traditional scientific

---

[2] The explanation is made also with referring the homepage by Ward, P.A.: http:// optimumsportsperformance.com/blog/data-analysis-template-in-r-markdown-jupyter-notebook/.

research. Researchers adopt this process when the problems that they would like to/should solve arise in the context of previous research, discussion with other researchers, or the requirement of non-academic problems that they are concerned with. Then, the researchers plan how to collect data, collect data through experiments and/or field research, and then analyze the data. When the hypotheses are validated by analysis, they become new knowledge. After confirming this knowledge through a review process, it was published and shared. The PPDAC model is a formalization of this research process. Another type of research can be conducted in the field of KDD. Data are collected using methods such as sensors, and a hypothesis is constructed by observing the data. The starting point is data that are not planned. In some cases, researchers go back to planning and defining the problem, and then the analysis is started. In other cases, researchers once tried to analyze the data, and from the results, discovered some hypotheses that have already been validated with the data. In both cases, the hypotheses are constructed after the data are collected; therefore, we call this type of research *hypothesis generation*.

In the projects presented in Sect. 7.2, trajectory data mining is a hypothesis generation method because no hypotheses were generated before data collection. On the other hand, syllabus integration is hypothesis validation because the project is started on the hypothesis that data science education in Japanese universities is not very popular.

7.4 Manipulating Undesigned Data

7.4.1 Designed Data and Undesigned Data

Generally, design requires some goals. In the PPDAC model, the data used to solve the problem defined in the problem step are designed in the plan step. If the analysis methods are already chosen in this step, the goal of designing the data collected in the data step is somewhat clear. The data must be accepted by the analyzing methods as their inputs, and the output results must be interpreted as a solution to the problem and work as new knowledge after the conclusion step. When the collected data are stored in database systems, more goals might arise for the design from the viewpoint of database engineering. For example, if the data are relational, we must avoid anomalies in the data, which is one of the main subjects in relational database theory [15]. In theory, the dependency of attributes in the relational schemata is used to avoid anomalies. The data used in the hill-climbing model are not well designed. They are obtained by various methods or used for secondary purposes other than primary purposes. In the trajectory data mining project, the trajectory data were obtained by methods using mobile phones and GPS and recorded in the form of a table. The purpose of collecting data was unclear, and how to use them was not well designed. In syllabus data integration, the homepages of syllabi vary from university to university, because each of them is for the convenience of the people concerning the university, including students and teachers. No other purposes are assumed, and even the views of the homepages in browsers are not uniform, and the source HTML files vary not uniformly.

7.4.2 Undefined Data as Words in Mathematical Logic

Based on the experience of the two projects introduced in Sect. 7.2, we propose to regard undefined data as *words* (*symbols* or *terms* in mathematical logic). We follow the idea of mathematical logic [1, 8] and formal language theory [5], and define a language as a set of words or *sentences* which is a parsed sequence of words. The meaning of each word is separate from the word itself. In mathematical logic, the meaning given by the person who spoke or wrote the word is the *intended interpretation*. Quite another interpretation of the same sentence was allowed; therefore, manipulating sentences without their meaning was investigated. Mathematically, let d be a datum as a word and o be an object in the domain D that we are treating. Thus, the interpretation is a mapping

$$I : d \longrightarrow o.$$

In the trajectory data mining project, each pair (x, y) of latitude and longitude values in the data is a word, and the address a of the location (x, y) is its meaning. The address a *sometimes contains* more information about the place by combining other data from tour guide sites on the internet. Some address names have geometrical or historical meaning. This can enrich information from the data, and the results output by the mining algorithm can be easily interpreted. In the example above, we can see that a large highway interchange occurs in Tsuru and that there is a shrine as a famous sightseeing place at Jingu-mae by consulting tour guide sites; therefore, we can conclude that tourists who visited both rode a bus or drove a car for sightseeing. If we would like to know the length of the trajectory, the address a would not work, and the meaning (x, y) should be itself a vector of real numbers. In the syllabus integration project, the HTML document p of every syllabus homepage is a word, and its meaning is a tuple t generated from the document in the integrated data. The HTML language was originally designed to represent information in browsers so that humans can understand it; therefore HTML documents are not suitable for other purposes, including query answering. When we use documents for such purposes, we must capture their meanings and represent them in words in other languages. More precisely, the tuple t is a word representing the same syllabus as the original HTML document d. If we are concerned about the jobs of professors, we should generate another meaning from p.

7.4.3 Designing an Interpretation of Data

From the experience of the two projects, we conclude that what we have to design is the interpretation of data, not the design of the data themselves, when the data are undesigned. We can analyze the trajectory database by assigning a meaning a to each datum (x, y), and construct a syllabus database by assigning a meaning t to each homepage p. The problem is how to find the domain for interpreting data as words, and how to assign a meaning to every datum. Each datum has two features. The first is the attribute of an object that we would like to represent using the datum. Even if a datum is generated without a specific purpose, it represents some attributes of an object. A pair (x, y) of real numbers is generated by representing an object location p, the first value strands for its "latitude" and the second for its "longitude". If the pair is generated by measuring the body of a human, the first value is his/her height and the second is his/her weight. Here we refer to the attribute name as the *role* of the data. However, this role may not be a singleton. In the syllabus integration, we created several roles for an HTML document, and the role of data is defined as a list of attribute names. The second feature is the type of data, that is, the set of operations that can be applied to the datum. The datum (x, y) for a location is a pair of real numbers both of which are measured in degrees. If the datum is for a human body, x is a positive real number measured in centimeters and y is measured in kilograms. Therefore the feature of a designed datum d is a pair of its roles and types, where each type corresponds to an attribute in the roles. We let role (d) be the list of attributes for the roles of d. Then, the process of designing interpretation of undersigned data d consists of three steps:

- searching some word w related to the attribute names in role (d),
- searching another data e such that role (e) contains w, and
- define an interpretation by joining d and e.

The first step was to collect words related to the attribute name in role (d), assuming that every attribute name should be a word in a natural language. In some fields using designed data, role (d) is also designed and shared by people related to the fields. For designers of interpretation, it is important to find good words related to the attributes in role (d). Sometimes, homonymies must be considered. We would like to thank some dictionaries and thesauri. Knowledge Graphs [7] are useful tools for this step. Then, we need to search for data related to the words found in the previous step. We have to take into account the situation where many words may be listed, the found data may be useless, and therefore, backtrack to the previous step may occur and a loop may arise. Graphical representation of the relations among attributes will be helpful and should be developed, similar to knowledge graphs. The last step will not be implemented with the join operation for databases, and a new program must be developed, such as a syllabus integration project. Machine learning using neural networks is beneficial.

7.5   Improvement of the KDD Processes

In the field of software engineering, we can find a pair similar to that of hill-climbing and cyclic models. A well-known model for developing large-scale software is the *waterfall* model [11, 13]. The model consisted of seven steps: system requirements, software requirements, analysis, program design, coding, testing, and operations. This model represents the top-down manner of developing software, and the seven steps are illustrated as six continuous waterfalls, from the top-left to the bottom-right in [11, 13]. In this model, the next stage is not started until the previous stage is completed, and parallel work in different stages is generally not allowed (Fig. 7.3a). The other model is agile development, where the entire project is divided into small units, and the planning, design, implementation, and testing are repeated many times in each unit. The agile model is typically illustrated as a cycle. Cycles for different units can operate simultaneously and sometimes without synchronization. In recent years, developments based on the agile model have become mainstream in software development (Fig. 7.3b).

Note that the waterfall model is not limited to software but has been adopted in various fields such as architecture and machine building when constructing large-scale artifacts. An example of a waterfall model for building construction consists of steps: requirements, preliminary design, construction design, construction, inspection, and completion. No backtracks were allowed in the original waterfall model. Under the strict interpretation of this constraint, the entire process must be stopped if some problems are found in an intermediate step in a process, and a great cost would be lost. The original study [11] proposed that an unconstrained backtrack must be prohibited (Fig. 7.4a) but a well-constrained backtrack should be allowed (Fig. 7.4b). For example, backtracking from one step to another should be allowed. This can be observed in the building construction process. The process is irreversible after the design is fixed and construction starts. Therefore, the people who meet the requirements and designers must be deeply discussed, and the designers and constructors

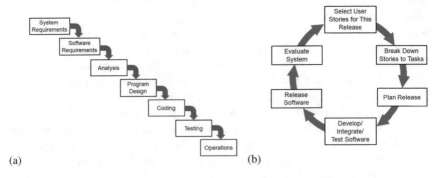

(a)   (b)

Fig. 7.3   Two Software Development Process models, the waterfall **a** and the extreme programming cycle **b**

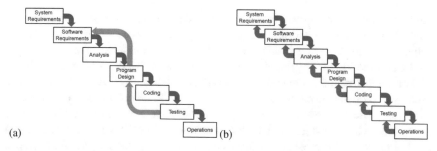

Fig. 7.4 Backtracks in the waterfall model, prohibited **a** and allowed **b**

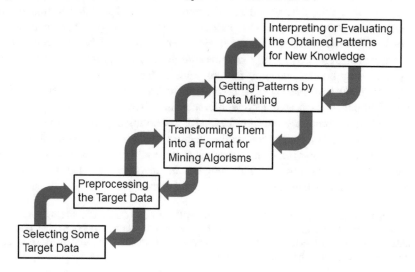

Fig. 7.5 Possible backtracks in the hill-climbing model

(carpenters) must communicate with each other. This results in partial loops with one-or two-step backtracks. The discussion of the interpretation of undersigned data in the previous section shows a loop between the first step, selecting data, and the second step, preprocessing the data (Fig. 7.5). This loop was not indicated in the original paper [2].

7.6   Conclusion and Remarks

In this article, we point out the importance of treating undesigned data to discover knowledge. The category of undesigned data is clarified by re-visiting KDD process models, and with the role of hypotheses in a model. We propose the treatment of undesigned data by referring to mathematical logic. We also show that the process model of KDD can be refined by referring to the process models of fields other than KDD. In [12], a researcher of psychology analyzed the process of his own research activity and summarized it as another model, which is a mixture of the hill-climbing model and the PPDAC cycle. He is not satisfied with the PPDAC because, in his research area, researchers must open all the data that they used in completing papers. Moreover, only data used in research activities should be opened before completed papers are reviewed, so that the data might be important for other research, even if the papers are not highly evaluated. This does not mean that the PPDAC cycle is becoming old-fashioned, but the role of open data has become more attractive. When using open data in the same research field, they should be well-designed, but when they are used in quite another area, they would be undesigned and the discussion in this article would be needed. A similar pair of hypothesis validations and hypothesis generation can be found in the field of automated theorem proving [1, 8]. Normally, automated theorem proving refers to proving theorems with computer programs, where theorems are provided by human beings to the programs as their inputs. Proving theorems are typical deductive inferences, and some AI researchers try to implement other types of inference, such as inductive inference and abductive inference [3]. In the implementation, some theorems are needed for generating hypotheses; therefore, an automated mechanism for generating the theorem was developed. As a variation of automated theorem proving, some researchers have investigated systems that help humans prove theorems [4, 10]. This suggests that discovery assistant systems will appear as an intermediate between the two process models.

Acknowledgements The authors would like to thank Prof. Yukio Ohsawa for his kind invitation to submit this article. They are also thankful to Siqi Peng and Kazuki Kawamura for their deriving results, which are the basis of the discussion in this article. We thank Atsuhiro Uematsu for explaining the process model used in the field of architecture.

References

1. Chang, C.L., Lee, R.C.T.: Symbolic Logic and Mechanical Theorem Proving. Academic Press (1973)
2. Fayyad, U., Piatetsky-Shapiro, G., Smyth, P.: From data mining to knowledge discovery in databases. AI Mag. **17**(3), 37–54 (1996)
3. Flach, P.A., Hadjiantonis, A.: Abduction and Induction: Essays on their Relation and Integration, vol. 18. Springer (2000)
4. Geuvers, H.: Proof assistants: history, ideas and future. Sadhana **34**(1), 3–25 (2009)
5. Hopcroft, J.E., Motwani R., Ullman, J.D.: Introduction to Automata Theory, Languages, and Computation (2007)
6. Kawamura, K., Yamamoto, A.: HTML-LSTM: information extraction from html tables in web pages using tree-structured LSTM. In: International Conference on Discovery Science, pp. 29–43. Springer (2021)
7. Kejriwal, M., Knoblock, C.A., Szekely, P.: Knowledge Graphs: Fundamentals, Techniques, and Applications. MIT Press (2021)
8. Loveland, D.W.: Automated Theorem Proving: A Logical Basis (1978)

9. Peng, S., Yamamoto, A.: Mining disjoint sequential pattern pairs from tourist trajectory data. In: International Conference on Discovery Science, pp. 645–658. Springer (2020)
10. Pfenning, F.: Logical framework. In: Handbook of Automated Reasoning (2001)
11. Royce, W.W.: Managing the development of large software systems. In: Proceedings of IEEE WESCON, pp. 1–9 (1970)
12. Saito, S.: Research culture and open practice in psychology (in Japanese). J. Jpn. Soc. Inf. Knowl. **31**(4), 446–451 (2021)
13. Sommerville, I.: Software Engineering, 10th edn. (2015)
14. Spiegelhalter, D.: The Art of Statistics: Learning from Data (2019)
15. Ullman, J.: Principles of Databases and Knowledge-Bases. vol. 1 and 2 (1988)

Chapter 8
Why Can We Obtain Such an Analysis?

Akinori Abe

Abstract Recently we have been able to dealt with very big data. Accordingly, it is possible to obtain results by Deep Learning in a certain quality. However, it is not good or, is dangerous to use the results blindly. For instance, by Google some persons pictures had automatically been labelled as golliras. Accordingly recently explainable deep learning has been proposed. It may be able to ask the system "why such a result is obtained?" by the method. Moreover, it will be better to explain the result logically. In this paper, we will propose an abductive method to explain the results for decision making.

Keywords Big data · Machine learning · Abduction · Analogical reasoning · Analogy

8.1 Introduction

Recently we have been able to dealt with very big data. Amoore and Piotukh pointed out that "[t]he twenty-first-century rise of big data marks a significant break with conventional statistical notions of what is *of interest*. The vast expansion of unstructured digital data, much of it considered to be open source, has been closely intertwined with the development of advanced analytical algorithms to make some sense of that data. And so, amid the cacophony of noise around the big in big data, we urge careful attentiveness to the work of the little analytics. Rather as the growth of statistical probabilistic methods made data on murder, health, employment or war perceptible and amenable to analysis, so contemporary analytics are instruments of perception without which the extensity of big data would not be perceptible at all" [7].

That is, thanks to (open) big data, it is possible by using statistical probabilistic methods to be perceptible and amenable to analyze the data to obtain a certain visible knowledge.

They continue: "First, the advent of advanced analytics ushers in a specific and novel *epistemology* of population. At first glance, the formulation $n = all$ appears to

A. Abe (✉)
Chiba University, 1-33 Yayoi-cho, Inage-ku, Chiba 263-8522, Japan
e-mail: ave@chiba-u.jp; ave@ultimaVI.arc.net.my

© Springer Nature Switzerland AG 2023
Y. Ohsawa (ed.), *Living Beyond Data*, Intelligent Systems Reference Library 230,
https://doi.org/10.1007/978-3-031-11593-6_8

render the whole of population as the sample—all data on all of life's transactions are, at least in theory, available to analysis. But the population as the sample in contemporary analytics processes does not imagine the population as a 'curve of normality' or a Gaussian bell curve of plotted attributes. Once we 'give up on the magician's wand' and 'follow the process,' as advised by Bergson, we can see how the object of interest becomes detached from the population as such. As the mathematicians and analysts tell us, advanced analytics work not merely with a statistical notion of what is interesting, but also via an inductive process of knowledge discovery, in which the process generates the rules. [...] Second, the advent of data analytics brings significant ontological implications for thought and practice. The processes of ingestion, partitioning and machinic memory reduce heterogeneous forms of life and data to homogenous spaces of calculation. Algorithmic technologies, such as those we have described, tend to reduce differences in kind to differences in degree, as Bergson and Deleuze might say, or to distances data-point to data-point. From these processes of reduction and flattening it is thought that different kinds of life stories—patterns of life—emerge, and that interventions and decisions can be made on their basis. The affective world of an epilepsy sufferer, a sub-prime borrower, a border crosser, a terrorist or a criminal is thought to be excavable from the seams and joins of multiple data sources. But, what kinds of stories can be told with analytics? What happens to the things that cannot be spoken, or that which is not fully accessible to us even of ourselves? [...] in order to decide on what or who is interesting, and this can only ever be retroactive. Finally, the rise of analytics has important consequences for the form of contemporary politics. In effect, the analytical processes of ingestion, partitioning and reassembly and memory we have described make a particular claim in the world, they say $n = all$, this is the world, here it is, all data is rendered tractable. They carve out and convert radical heterogeneity into flat difference of degree, such that it appears as though everything is calculable, everything about the uncertain future is nonetheless decidable. [...] Perhaps the most striking and troubling matter of scale in big data is the recognition of an extensity that always exceeds the capacity of human knowledge to collect and apprehend it. The promise and allure of the little analytics is to see those things that would otherwise be invisible, to perceive the imperceptible and to feed the insights to those who would action them. Confronted with this claim to reduce the fallibility of governing and decision, to do so in a manner that is 'fault tolerant,' a political response must be mindful of the perils of the magician's wand, and point to the material, contingent and fallible processes that make this claim possible."

In this way, they very seriously pointed out the problem. First, data usually are not normally distributed, especially if the data set is not sufficiently big. In general, systems of analysis assume that the data are normally distributed. Second, during analysis, the construction of data will change and the meaning of data might change or be misunderstood. A more serious problem they point out is the troubling matter of scale in big data is the recognition of an extensity that always exceeds the capacity of human knowledge to collect and apprehend the changing data. This type of problem is the so-called frame problem. It will be beneficial to access big data, but it is not easy to deal with the data in a satisfactory manner.

Regarding machine learning, Athey pointed out that "[a] recent explosion of analysis in science, industry, and government seeks to use ' big data' for a variety of problems. Increasingly, big-data applications make use of the toolbox from supervised machine learning (SML), in which software programs take as input training data sets and estimate or 'learn' parameters that can be used to make predictions on new data. In describing the potential of SML for clinical medicine, Obermeyer et al. have commented that 'Machine learning... approaches problems as a doctor progressing through residency might: by learning rules from data. Starting with patient-level observations, algorithms sift through vast numbers of variables, looking for combinations that reliably predict outcomes. Where machine learning shines is in handling enormous numbers of predictors—sometimes, remarkably, more predictors than observations—and combining them in nonlinear and highly interactive ways.' SML techniques emerged primarily from computer science and engineering, and they have diffused widely in engineering applications such as search engines and image classification. More recently, the number of applications of SML to scientific and policy problems outside of computer science and engineering has grown. In the public sector, SML models have been introduced in criminal justice; for predicting economic well-being at a granular level using mobile data, satellite imagery, or Google Street View; and for allocating fire and health inspectors in cities, as well as a variety of other urban applications. The techniques have also been used to classify the political bias of text or the sentiment of reviews. In medicine, SML-based predictive algorithms have been implemented in hospitals to prioritize patients for medical interventions based on their predicted risk of complications, and in a wide variety of additional medical applications, including personalized medicine" [8]. She then concluded that "[o]verall, for big data to achieve its full potential in business, science, and policy, multidisciplinary approaches are needed that build on new computational algorithms from the SML literature, but also that bring in the methods and practical learning from decades of multidisciplinary research using empirical evidence to inform policy. A nascent but rapidly growing body of research takes this approach: For example, the International Conference on Machine Learning (ICML) in 2016 held separate workshops on causal inference, interpretability, and reliability of SML methods, while multidisciplinary research teams at Google, Facebook, and Microsoft have made available toolkits with scalable algorithms for causal inference, experimental design, and the estimation of optimal resource allocation policies. As the SML research community and other disciplines continue to join together in pursuit of solutions to real-world policy problems using big data, we expect that there will be even greater opportunities for methodological advances, as well as successful implementations, of data-driven policy."

Perhaps there exist several problems in dealing with big data. For instance, the data might not be normally distributed, and it is in addition rather difficult to manage big data. Nonetheless, we are very happy to have access to very huge data in the world to obtain a certain body of knowledge (as a result) by applying, for instance, Deep Learning to gain results with a certain satisfactory quality. However, it is not good, and indeed can be dangerous to use the results blindly. For instance, problems have arisen from doing so. In 2015, Google's new Photos app categorized photos in one of the most racist ways possible. On June 28th, computer programmer Jacky Alcine found that the feature kept tagging pictures of him and his girlfriend as "gorillas." He tweeted Google asking what kind of sample images the company had used that would allow such a terrible mistake to happen [11]. It has been pointed out that "[f]or years, Kodak used a coating on its film that favored Caucasian skin tones, making it more difficult to shoot darker skin. Nikon and other consumer camera companies have also had a history of showing bias to white faces with their facial recognition software." Thus Zunger (Google's chief social architect) said that Google has had similar issues with facial recognition due to inadequate analysis of skin tones and lighting. Google attempted to fix the algorithm, but ultimately removed the gorilla label altogether.

In addition, Rudin pointed out that "[m]any of the ML models are black boxes that do not explain their predictions in a way that humans can understand. The lack of transparency and accountability of predictive models can have (and has already had) severe consequences..." [19].

In this chapter, I point out the problems in the results of machine learning. As a machine learning system is a type of a black box, as a result the obtained result cannot readily (or ever) be explained. I will then demonstrate a certain method to explain why can we obtain such an analysis by abduction and analogical reasoning.

8.2 (Big) Data Society

Today we are said to live in a big data society. Through linked data (semantic web) technology, several sets of data are linked and combined as a big data set. Amoore and Piotukh pointed out that "[t]here can be little doubt that the very idea of big data is having significant consequences for economy and society, and for human knowledge— whether in the petabytes of scientific data generated by the Large Hadron Collider at CERN, or in the ongoing debates on the social sciences' use of transactional data for the understanding of human behaviour and social transformation. The widely held view that we are living in a world dominated by the 4 Vs of big data—increased volume, variety, velocity, veracity—has led to a focus on the significance of the scale and scope of the digital traces left in the wake of people, things, money and ideas on the move. As data-generating devices proliferate, and as data storage and processing power have become more scalable, the rise of twenty-first-century big data has been described as a 'goldmine' of 'magical material,' a 'new oil' fuelling innovative forms of economic transaction and circulation whose 'core assets' are data" [7].

Accordingly, it will be necessary/better to utilize these big data for decision making. Usually machine learning such as deep learning is used to analyze data. Deep learning systems seem to generate suitable results.

However, several severe problems have arisen regarding this. In 2015, Google's new Photos app categorized photos in one of the most racist ways possible. On June 28th, computer programmer Jacky Alcine found that the feature kept tagging pictures of him and his girlfriend as "gorillas." He tweeted Google, asking what kind of sample images the company had used that would allow such a terrible mistake to happen

[11]. It has in fact been pointed out that "[f]or years, Kodak used a coating on its film that favored Caucasian skin tones, making it more difficult to shoot darker skin. Nikon and other consumer camera companies have also had a history of showing bias to white faces with their facial recognition software." Thus Zunger (Google's chief social architect) said that Google has had similar issues with facial recognition due to inadequate analysis of skin tones and lighting. Google attempted to fix the algorithm, but ultimately it removed the gorilla label altogether.

What is the problem? As Rudin pointed out [19], "[m]any of the ML models are black boxes that do not explain their predictions in a way that humans can understand. The lack of transparency and accountability of predictive models can have (and has already had) severe consequences. That is, the problem is (can) not to explain why the result was generated."

Then Rudin continued that, "[r]ather than trying to create models that are inherently interpretable, there has been a recent explosion of work on 'Explainable ML,' where a second (posthoc) model is created to explain the first black box model. This is problematic. Explanations are often not reliable and can be misleading, as we discuss below. If we instead use models that are inherently interpretable, they provide their own explanations, which are faithful to what the model actually computes."

Thus, the most serious problem is that a certain rule can easily be obtained by computational learning such as deep learning, but we do not check the result or the process by which the result was obtained. That is, the result is usually used blindly without criticism. Furthermore, since the learning process is a black box system, it is impossible to explain how the result was generated. This is the most serious problem.

8.3 Treatment of Data in the World

For the problems pointed out in the previous sections, some proposals have been proposed recently.

I think it is important to keep the interpretability when we show a certain result.

Rudin defined interpretability thus: " [i]nterpretability is a domain-specific notion, so there cannot be an all-purpose definition. Usually, however, an interpretable machine learning model is constrained in model form so that it is either useful to someone, or obeys structural knowledge of the domain, such as monotonicity, causality, structural (generative) constraints, additivity, or physical constraints that come from domain knowledge. Interpretable models could use case-based reasoning for complex domains. Often for structured data, sparsity is a useful measure of interpretability, since humans can handle at most 7 ± 2 cognitive entities at once. Sparse models allow a view of how variables interact jointly rather than individually. We will discuss several forms of interpretable machine learning models for different applications below, but there can never be a single definition; e.g., in some domains, sparsity is useful, and in others is it not. There is a spectrum between fully transparent models (where we understand how all the variables are jointly related to each other) and models that are lightly constrained in model form (such as models that are forced to increase as one of the variables increases, or models that, all else being equal, prefer variables that domain experts have identified as important, [...])" [19].

The important point is that an interpretable machine learning model is constrained in model form so that it is either useful to someone, or obeys structural knowledge of the domain, such as monotonicity, causality, structural (generative) constraints, additivity, or physical constraints that come from domain knowledge.

Rudin continued, "[e]xplanations must be wrong. They cannot have perfect fidelity with respect to the original model. If the explanation was completely faithful to what the original model computes, the explanation would equal the original model, and one would not need the original model in the first place, only the explanation. (In other words, this is a case where the original model would be interpretable.) This leads to the danger that any explanation method for a black box model can be an inaccurate representation of the original model in parts of the feature space. An inaccurate (low-fidelity) explanation model limits trust in the explanation, and by extension, trust in the black box that it is trying to explain. An explainable model that has a 90% agreement with the original model indeed explains the original model most of the time. However, an explanation model that is correct 90% of the time is wrong 10% of the time. If a tenth of the explanations are incorrect, one cannot trust the explanations, and thus one cannot trust the original black box. If we cannot know for certain whether our explanation is correct, we cannot know whether to trust either the explanation or the original model [19].

Interpretability will be based on explanations. Thus, for computer-generated results, interpretability is very important, but it is more important that the explanation be fully transparent and correct. The quality of interpretability depends upon the correct explanation, and it is difficult to arrive at a correct explanation only by using a black box model.

8.4 Abduction and Analogy

In this section, I will briefly illustrate abduction and show the mechanism of hypothetical reasoning (Theorist), which can be regarded as a computational abduction, and Abductive Analogical Reasoning, which is an extension of abduction.

8.4.1 Philosophical Abduction

Peirce classified *abduction* from a philosophical point of view as the operation of adopting an explanatory hypothesis and characterized its form [14].

(1) The surprising fact, C, is observed;
(2) But if A were true, C would be a matter of course,
(3) Hence, there is reason to suspect that A is true.

Peirce then illustrated abduction as follows:

...abduction is an operation for adopting an explanatory hypothesis, which is subject to certain conditions, and that in pure abduction, there can never be justification for accepting the hypothesis other than through interrogation.

The important keyword of the definition is, *explanatory hypothesis*. That is, abduction is performed as an explanation of something. Peirce would not assume the performance of computational abduction.

8.4.2   Computational Abduction

The first achievement of computational abduction will beas proposed by [16]. Pople explored the abductive reasoning process and developed a model for its mechanization, which consists of an embedding of deductive logic in an iterative hypothesis and test procedure. After several years, several more logical abduction systems such as Theorist [15] and ALP (Abductive Logic Programming) [12] weare proposed.

Theorist is considered as a computational hypothetical reasoning system. HA hypothetical reasoning, which is an explanatory form of reasoning, generates (collects, selects) a consistent hypothesis set from hypothesis candidate hypothesess to explain the given observation. The generated hypothesis set can be regarded as an answer (solution) which that can explain the observation. The inference mechanism of Theorist is described as follows:

$$F \nvdash O. \qquad (O \ cannot \ be \ explained \ only \ by \ F.) \qquad (8.1)$$

$$F \cup h \vdash O. \qquad (O \ can \ be \ explained \ by \ F \ and \ h.) \qquad (8.2)$$

$$F \cup h \nvdash \square. \qquad (F \ and \ h \ are \ consistent.) \qquad (8.3)$$

where F is called a fact, which is always true. On the other hand conversely, h is called a hypothesis, which is not always true and is included in the hypothesis base H ($h \subseteq H$). O is an observation to be explained. $\square$ is an empty set. When $F \cup h \vdash \square$, F and h are not consistent. In the knowledge base, inconsistent information is also included in the knowledge base.

The limitation of this abduction is that, a hypothesis base should be prepared.

8.4.3   The Clause Management System (CMS)

The Clause Management System (CMS), was proposed by Reiter and de Kleer [18], and it was a database management system. Its mechanism is illustrated as follows:

When $\Sigma \not\models C$, if propositional clause C (observation) is given, CMS returns a set of minimal clauses S to clause set Σ such that

$$\Sigma \models S \vee C. \tag{8.4}$$

$$\Sigma \not\models S. \tag{8.5}$$

A clause S is called a minimal support clause, and $\neg S$ is a clause set that is missing from clause set Σ that can explain C. Therefore, although CMS was not proposed as abduction, since from the abductive point of view $\neg S$ can be thought of as an abductive hypothesis, CMS can be used for abduction.

8.4.4 Analogical Reasoning

In https://thedecisionlab.com/reference-guide/psychology/analogy/, it is pointed out that "[c]omplex phenomena can be difficult to grasp. We often refer to simpler concepts to make this complexity intelligible. For example, try to imagine an atom's structure? it's impossible because of its microscopic size! And it's difficult to comprehend something we can't imagine because we lack a reference point. To picture an atom, you might compare it to the solar system. The nucleus is the sun and the orbiting planets the electrons and neutrons. In this instance, we used the solar system as a source in an analogy with our target, atomic structure, to enhance our understanding."

The similarity is used to determine unknown phenomena in the unknown field. Actually the result may not fully correct, but it will be a certain hint. The mechanisms of analogical learning wereas discussed in [10].

In the computational analogy, a the structural analogy is usually applied to. However, simple similarity can be also used. For instance, a concept base [13] can measure the similarity between objects. Actually In fact, it does not adopt structure, but considers a viewpoint (context). By this score analogical reasoning can be achieved.

8.4.5 Abductive Analogical Reasoning (AAR)

The Clause Management System (CMS) shown above generates only the minimal hypothesis set. Thus, it is not always the case that we can obtain the a sufficient hypothesis set. Accordingly I proposed Abductive Analogical Reasoning (AAR) [2], which that logically and analogically generates missing hypotheses. Its generation mechanism is similar to that of CMS's. Structures of generated knowledge sets are analogous to the known knowledge sets. In the framework of AAR, not completely unknown but rather unknown hypotheses can be generated. In addition, by the introduction of analogical mapping, we can adopt new hypothesis evaluation criteria other than besides Occam's Razor (for instance, criteria such as explanatory coherence [20]). The inference mechanism is briefly illustrated as follows (for notations, see [2]):

When

$$\Sigma \not\models O, \qquad (O \text{ cannot only be explained by } \Sigma.) \qquad (8.6)$$

Σ (background knowledge) lacks a certain set of clauses to explain O. Consequently, AAR returns a set of minimal clauses S such that

$$\Sigma \models S \lor O, \qquad (8.7)$$

$$\neg S \notin \Sigma. \qquad (8.8)$$

The result is the same as that of CMS's. This is not always a guaranteed hypothesis set. To guarantee the hypothesis set, we introduced analogical mapping from known knowledge sets.

$$S \mapsto S', \qquad (S' \text{ is analogically transformed from } S.) \qquad (8.9)$$

$$\neg S' \in \Sigma, \qquad (8.10)$$

$$S' \mapsto S'', \qquad (8.11)$$

$$\Sigma \models S'' \lor O, \qquad (8.12)$$

$$\neg S'' \notin \Sigma. \qquad (8.13)$$

O is then explained by $\neg S''$ as an hypotheses set. Thus, we can generate a new hypothesis set that is logically abduced whose structure is similar to authorized (well-known) knowledge sets.

8.5 Data Managemant by Abduction and Analogy

In the previous sections, I pointed out that the interpretability of computer-generated results is very important, but it is more important that the explanation be fully transparent and correct. The quality of interpretability depends upon the correct explanation; it is difficult to arrive at the correct explanation using only a black box model.

As Rudin pointed out [19], "[m]any of the ML models are black boxes that do not explain their predictions in a way that humans can understand. The lack of transparency and accountability of predictive models can have (and has already had) severe consequences. Recently many researchers have tried to attack this problem." However, Rudin continues, "[r]ather than trying to create models that are inherently interpretable, there has been a recent explosion of work on 'Explainable ML,' where a second (posthoc) model is created to explain the first black box model. This is problematic. Explanations are often not reliable, and can be misleading, as we discuss below. If we instead use models that are inherently interpretable, they provide their own explanations, which are faithful to what the model actually computes."

In this section, I will show the abductive and analogical strategy to solve the above problem.

A serious problem that Google faced was the categorization of people as "gorillas"; Google attempted to fix the algorithm, but ultimately removed the gorilla label altogether. The second (posthoc) model may be wrong, because the model will not be created logically. The procedure is a black box. With a logical procedure, a reliable model will be created whose process is not a black box, allowing us to see how the result is generated.

In [3], I mentioned that in the Innovation Game, participants do their best to explain a certain requirement by generating or collecting a set of hypotheses (in this case, a set of techniques) in the current environment. Many creative activities can be performed as a case of abduction. This type of inference strategy can be considered a form of abduction. In that study I showed the process of making a decision; here I quote a part of the strategy.

For instance, requirements such as "I would like to take time off work to go see the Olympic game" were given. A proposal then given was, "A fake tweet can be shown in an open account."

Techniques to achieve the above proposal included "shop arrival count data." The theme of the games was how to make the 2020 Olympic games successful. Original techniques were extracted from the application to solve problems in university students' lives. Thus, the games was conducted using techniques from a rather different theme and field. A participant who proposed this interpreted the requirement as "going to see the Olympic games by faking his/her place." He/she then generates a technique satisfying the requirements to pretend his/her place and to make use of a tweet technique (perhaps of the Olympic games) by manipulating "shop arrival count data" rather differently from the original usage.

That is, according to the following formula, it is possible to generate a technique set to satisfy the requirements.

$$\textit{knowledge of the place where the Olympic games are being held} \cup$$
$$\{\textit{show a different place} \cup \textit{perform tweet}\} \vdash$$
$$\textit{pretend his/her place.} \quad (8.14)$$

Basically, in hypothetical reasoning, the inference is performed by the selection of the existing hypothesis set from the hypothesis base or by the unification of variables in the hypothesis set. However, in the above example, the techniques are not used as they are, and the requirement is interpreted differently. Hypothetical reasoning basically means that we use the hypotheses as they are for inference. If we would like to produce what we want to really have, it is sometimes impossible to use only the existing techniques to make a brand new product. The above example shows a similar situation. When we make a brand new product, we naturally extend the possible techniques. The situation can be illustrated by the framework of abduction.

In the previous section, CMS was illustrated, which can generate a set of hypotheses necessary for inference but missing from the hypothesis base. In addition, AAR is illustrated to be an extension of CMS, which introduces a similar hypothesis set generation.

By using CMS and AAR, the above strategy can be explained. For instance, in the above example, suppose we do not have the technique "perform tweet." Then

$$\textit{knowledge of the place where the Olympic games are being held} \cup$$
$$\{\textit{show a different place}\} \not\vdash \textit{pretend his/her place.} \quad (8.15)$$

If we follow CMS,

$$\textit{knowledge of the place where the Olympic games are being held} \not\models$$
$$\{\neg\textit{show a different place}\} \vee \textit{pretend his/her place.} \quad (8.16)$$

can be obtained. In order to satisfy the requirement, it is necessary to find a minimal S supporting the following formula.

$$\textit{knowledge of the place where the Olympic games are being held} \models$$
$$S \vee \{\neg\textit{show a different place}\} \vee \textit{pretend his/her place.} \quad (8.17)$$

If we can obtain $S = \neg tell$, the above formula can be satisfied.

$$\textit{knowledge of the place where the Olympic games are being held} \models$$
$$\{\neg tell \vee \neg\textit{show a different place}\} \vee$$
$$\textit{pretend his/her place.} \quad (8.18)$$

As for the clause "show a different place," a technique such as "shop arrival count data" can be used to achieve it.[1] The most difficult problem is how to generate a clause "tell." This is possible with the introduction of AAR to generate a clause "tell."

When the following formula is obtained;

$$\textit{knowledge of the place where the Olympic games are being held} \models$$
$$\{\neg tell \vee \neg\textit{shop arrival count data}\} \vee$$
$$\textit{pretend his/her place.} \quad (8.19)$$

---

[1] It is necessary to provide the necessary background knowledge to perform this.

it will be necessary to search or generate a technique similar to "tell." For instance, if we interpret "tell" as "communicate his/her voice," "broadcast" ("tell $\approx$ broadcast") can be searched from the knowledge base in another context or else newly generated. In addition, if we interpret "tell" as "communicate phrase (including characters)," "perform tweet" ("tell $\approx$ perform tweet") can be searched from the knowledge base in another context or newly generated. The selection of the hypothesis set is determined in consideration of whether the hypothesis set is consistent with "the place where the Olympic games are being held" and other techniques, and if it can be used in the context. If it is not consistent, it is necessary to produce new techniques or to replace it with other techniques. This is a non-monotonic aspect of such reasoning. In this case, if personal information is "broadcast" in "the place where the Olympic games are being held," it will reveal personal information. Therefore, "broadcast" is not consistent with the situation. "Perform tweet" has a rather pronounced publication feature, but it may not be so strong as for "broadcast." Accordingly, "perform tweet" can be adopted as a similar technique as "tell." The proposal shown in the KeyGraph sheet (please see [3]) is determined, because the proposer thought "perform tweet" was enough to maintain the privacy of the individual. If it is still insufficient, it will be necessary to adopt another technique that can better maintain the privacy of the individual, such as "send a message via FB (facebook)" ("tell $\approx$ send a message via FB"). This type of non-monotonic hypothesis selection continues until a consistent hypothesis set can be generated. In addition, when the generated hypothesis set cannot be used as it is, an analogical mapping is introduced to generate a new and usable hypothesis set in the situation.

$$knowledge\ of\ the\ place\ where\ the\ Olympic\ games\ are\ being\ held\ \mathrel{\mid\approx}$$
$$\{\neg send\ a\ message\ via\ FB \vee \neg shop\ arrival\ count\ data\}$$
$$\vee pretend\ his/her\ place \qquad (8.20)$$

AAR can generate a more complex hypothesis set than the minimal hypothesis set. Accordingly, the obtained hypothesis set can be more reasonable hypothesis set by referring to the other existing techniques. Of course, during analogical mapping, the consistency of the hypothesis set should be considered.

The above shows an abductive strategy to explain how to satisfy the given requirement. It can be seen that the process is very transparent, and the process of generating the result is very clear. We can determine how to obtain the results, an important feature that DNN lacks. This type of strategy, moreover, can be applied to check the problem mentioned previously of the case of Google to determine why "gorillas" was generated. The reason cannot be observed by learning. Perhaps the model focused only on the color type and hair type. Though Zunger (Google's chief social architect) said that "Google has had similar issues with facial recognition due to inadequate analysis of skin tones and lighting. We used to have a problem with people (of all races) being tagged as dogs," in fact no reasonable reason was shown.

In the case of this problem, it is necessary to explain why a person was identified as a gorilla. If the observation is of a gorilla, abduction can perhaps generate a reason as a hypothesis set that may or may not coincide with the model generated by learning. It is then possible to check the hypothesis.

Ras et al. [17] reviewed several explainable deep learning systems, mentioning that "[t]he goal of any explanation can roughly fall into one of the following two categories: (i) explanations that give insight into model training and generalization. These explanations give a practitioner additional information that can be used to make decisions about the components in the model training and validation process, e.g., the number of labeled data, value of the hyperparameters, model choice. The other category is (ii) explanations that give insight into model predictions. Most explanations fall into this category and help practitioners explain why the model made a particular prediction, usually in terms of the model input."

They categorized explainable DNNs under the following three categories:

- Visualization methods: Visualization methods express an explanation by high-lighting, through a scientific visualization, characteristics of an input that strongly influence the output of a DNN.
- Model distillation: Model distillation develops a separate, "white-box" machine learning model that is trained to mimic the input-output behavior of the DNN. The white-box model, which is inherently explainable, is meant to identify the decision rules or input features influencing DNN outputs.
- Intrinsic methods: Intrinsic methods are DNNs that have been specifically created to render an explanation along with its output. As a consequence of their design, intrinsically explainable deep networks can jointly optimize model performance and the quality of the explanations produced.

They examined what makes a good explanation, pointing out that the traits of fidelity, consistency, stability, and comprehensibility are most commonly scrutinized and discussed.

It is important to evaluate the generated explanation, and the evaluation will be conducted by human beings. The results generated by a black box, however, cannot be evaluated by computers because no process of model generation is shown.

The quality of interpretability will thus depend upon the correct explanation.

8.6 Conclusions

In this chapter, I discuss how to explain the process by which the results are generated.

I pointed out the problem using Google's gorilla incident: There was no transparent process to show how the results were generated. By using an abductive procedure, it is possible to show the logical process by which the results are generated.

References

1. Abe, A.: Two-sided hypotheses generation for abductive analogical reasoning. In: Proceedings of the ICTAI99, pp 145–152 (1999)
2. Abe, A.: Abductive analogical reasoning. Syst. Comput. Jpn. **31**(1), 11–19 (2000)
3. Abe, A.: The role of abduction in IMDJ. In: Proceedings of IJCAI2015 International Workshop on Chance Discovery, Data Synthesis and Data Market, pp. 59–64 (2015)
4. Abe, A.: Abduction dealing with potential values and its datasets towards IMDJ. Int. Decis. Technol. IOS Press **10**(3), 223–233 (2016)
5. Abe, A: On the logical and ontological treatment of IMDJ data. In: Proceedings of ECAI2016 2nd European Workshop on Chance Discovery and Data Synthesis, pp. 33–38 (2016)
6. Abe, A.: Interpretable AI as curation. In: Proceedings of AAAI Spring Symposia on Interpretable AI for Well-Being: Understanding Cognitive Bias and Social Embeddedness (2019)
7. Amoore, L., Piotukh, V.: Life beyond big data: governing with little analytics. Econ. Soc. **44**(3), 341–366 (2015)
8. Athey, S.: Beyond prediction: using big data for policy problems. Science **355**, 483–485 (2017)
9. Garratt, J.: The Gap Between Making A Prediction and Making A Decision (2020). https://livebeyonddatascience.com/the-gap-between-making-a-prediction-making-a-decision/
10. Gentner, D: The mechanisms of analogical learning. Similarity and Analogical Reasoning, pp. 199–241. Cambridge University Press (1989)
11. Grush, L.: Google engineer apologizes after Photos app tags two black people as gorillas (2015). https://www.theverge.com/2015/7/1/8880363/google-apologizes-photos-app-tags-two-black-people-gorillas
12. Kakas, A.C., Kowalski, R.A., Toni, F.: Abductive logic programming. J. Log. Comput. **2**(6), 719–770 (1992)
13. Kasahara, K., Matsuzawa, K., Ishikawa, T., Kawaoka, T.: Viewpoint-based measurement of semantic similarity between words. In: Proceedings of 5th International Workshop on Artificial Intelligence and Statistics, pp. 292–302 (1995)
14. Peirce, C.S.: Philosophical Writings of Peirce. Dover (1955)
15. Poole, D., Goebel, R., Aleliunas, R.: Theorist: a logical reasoning system for defaults and diagnosis. In: Cercone, N.J., McCalla, G. (eds.), The Knowledge Frontier: Essays in the Representation of Knowledge, pp. 331–352. Springer (1987)
16. Pople, H.E. Jr.: On The mechanization of abductive logic. In: Proceedings of IJCAI73, pp.147–152 (1973)
17. Ras, G., Xie, N., Dorany, D.: Explainable Deep Learning: A Field Guide for the Uninitiated (2021). arXiv:200414545v2 [cs.LG], https://arxiv.org/abs/2004.14545
18. Reiter, R., de Kleer, J.: Foundation of assumption-based truth maintenance systems: preliminary report. In: Proceedings of AAAI87, pp. 183–188 (1987)
19. Rudin, C.: Stop explaining black box machine learning models for high stakes decisions and use interpretable models instead. Nat. Mach. Intell. **1**(5), 206–215 (2019)
20. Thagard, P.: Explanatory coherence. Behav. Brain Sci. **12**, 435–502 (1989)

Part III
Insights from Cases

Chapter 9
Positive Artificial Intelligence Meets Affective Walkability

Stefania Bandini, Francesca Gasparini, and Flavio S. Correa da Silva

Abstract Urban design has followed different schools and traditions, among which we highlight the *smart cities* and the *human-centered cities* initiatives, which contrast in certain respects and are complementary in others, similar to how *traditional psychology* and *positive psychology* are contrasting and complementary, and to how *artificial intelligence* has followed contrasting and complementary development pathways along *automation of human tasks* and *augmentation of human capabilities*. In this chapter we explore these traditions to establish the concept of *positively intelligent neighborhoods*, as a convergence of *intelligent sensing* and *positive technologies* to design *human-centered neighborhoods* based on the notion of *affective walkability* with the support of *positively intelligent design agents*.

Keywords Affective walkability · Positive technologies · Human-centered design · Positively intelligent neighborhoods · Positively intelligent design agents

All authors have contributed equally to this work. This work was developed with partial support from FAPESP *MCTIC/CGI—Future Internet for Smart Cities* and the INCT *InterSCity—Smart https://www.overleaf.com/project/61de658e82ec4e3bb27a3987 Cities (Brazil)* and from *Fondazione Cariplo—Longevicity—Social Inclusion for the Elderly through Walkability (Italy)*.

S. Bandini (✉) · F. Gasparini
Universitá degli Studi di Milano-Bicocca, Milan, Italy
e-mail: stefania.bandini@unimib.it

F. Gasparini
e-mail: francesca.gasparini@unimib.it

S. Bandini
RCAST - The University of Tokyo, Bunkyo City, Tokyo, Japan

F. S. C. da Silva
University of Sao Paulo, Sao Paulo, Brazil
e-mail: fcs@usp.br

© Springer Nature Switzerland AG 2023
Y. Ohsawa (ed.), *Living Beyond Data*, Intelligent Systems Reference Library 230,
https://doi.org/10.1007/978-3-031-11593-6_9

9.1 Introduction

The works of Douglas Engelbart (1925–2013) and John McCarthy (1927–2011) were of fundamental importance for the development of computer technology. Engelbart focused on designing and engineering devices to augment human capabilities, so that humans could work collectively to solve complex problems and build a better world [31]. For his work, Engelbart is considered one of the fathers of the field of *human-machine interaction*, and an advocate of computer technologies as a means to promote human collaboration. McCarthy organized the *Dartmouth Conference* in 1956, in which the name *artificial intelligence* was coined to bring together initiatives to reconstruct intelligent behavior in artifacts with the ultimate purpose of relieving humans from the need to perform repetitive, strenuous, and/or potentially harmful activities. McCarthy is considered one of the fathers of *artificial intelligence*.

The works of these fathers of important areas in computer technology highlight two complementary approaches to build technologies that support human actions and interactions: a *positive* approach with a focus on desired human capabilities to be enhanced *in* humans *for the benefit of* humans, and a *negative* approach with a focus on undesired human tasks and traits to be *retracted from* humans, also *for the benefit of* humans. It resonates with—and furthers—arguments presented by Winograd [46], suggesting that the design of intelligent systems can be refined by accounting for both negative and positive approaches.

The appropriateness of adopting methodologies that encompass both negative and positive approaches to design has been considered in different domains. In *positive psychology*, similar arguments ground the proposition that a *negative* approach focusing on the treatment of undesired psychological hallmarks should be complemented by a *positive* approach focusing on strengthening desired potentialities. *Positive technologies* and *positive computing* are offsprings of positive psychology toward the development of artifacts to support the flourishing of desired human potentialities.

Similarly, we find contrasting but also complementary approaches to urban design characterized as *smart cities* [1], based on the perspective of *cities as machines* [19] and focusing on the automation of services for improved efficiency in urban structures; and *human-centered cities*, based on the view of citizens as the *raison d'être* of cities [19, 29, 30] and focusing on human well-being as the central driver for urban design while seeking organizations of space that can enable and support human flourishing.

In the present chapter we follow the convergence approach suggested by Winograd [46] to find ways to best put artificial intelligence, and specifically *intelligent sensing technologies*, at the service of *human-centered urban design*, focusing on the notion of *affective walkability* to characterize *positively intelligent neighborhoods*. In Sect. 9.2 we introduce the fundamental concepts related to *positive psychology, positive technologies, positive computing*, and *positive artificial intelligence* as bricks to build the concept of *positively intelligent design agents*, which are software agents based on machine learning and artificial intelligence in general, capable of providing designers, specifically urban designers, with support to design citizen-friendly urban spaces. In Sect. 9.3 we see how *intelligent sensing* can be used to capture and interpret physiological signals from individuals as indicators of affective states, and how these states can be linked to public spaces for pedestrians, leading to the concept of *affective walkability*, formulated in such way that it can be empirically assessed and correlated with features of urban landscapes. We conclude this chapter with a *call to action* in Sect. 9.4 to employ the proposed sensing technologies to collect global data about *affective walkability* in order to provide urban designers with well-grounded data to support the design of *human-centered neighborhoods*.

9.2 Positive Artificial Intelligence

Positive psychology was initially formulated in the early 90s as a complement to traditional psychology—i.e., psychological methods based on psychoanalysis and the treatment of undesired hallmarks—by focusing on well-being, happiness, and positivity [24]. It is grounded on principles from humanistic psychology, logotherapy, Jungian analysis, and Ancient Greek and Eastern philosophies [13, 14, 20, 35, 45]. Positive psychology is grounded on *PERMA* theory [39] and *Flow* theory [10]:

- *PERMA* is an acronym for *P*ositive emotions, *E*ngagement, *R*elationships, *M*eaning, and *A*chievement, which are identified as the five factors influencing the path toward a life of fulfillment and happiness.
- *Flow* is a state of awareness of inner and outer events that leads to pleasure and satisfaction. A *state of Flow* is reached and sustained via the dynamic balance of perceived challenges and skills, about which the individual must be informed through feedback about inner and outer events and progress toward self-determined goals.

Seligman and Csikszentmihalyi [40] proposed a *calculus of well-being* to clarify how negative (i.e., traditional) and positive psychology could be balanced. Resorting to a simplistic analogy with Newtonian mechanics, retraction of undesired psychological hallmarks can neutralize acceleration of a patient toward an unwanted direction in life, but actual movement in a new direction can only be achieved by adding energy to potentialities that can overcome inertia and lead to a desired direction. The calculus of well-being challenges a mechanistic view of human personality with the understanding that *double negation does not equal affirmation*: Not being negative does not entail that something is positive when it comes to human personality and emotions.

In positive psychology we also find the difference between *hedonic* and *eudaimonic* experiences as complementary forces that lead to fulfillment and happiness: Hedonic experiences relate to momentary pleasure, contentment, and satisfaction, are not self-sustained, and require frequent extrinsic reinforcement, whereas eudaimonic experiences relate to self-actualization of prosperity, blessedness, and happiness, are perennial or at least long-lasting, and are self-sustained. Resorting to the same analogy with Newtonian mechanics, hedonic experiences can change the direction of movement of oneself but cannot alter the friction that works to stop that movement in the long run, whereas eudaimonic experiences change the direction of movement *and* reduce friction toward zero, hence being more deeply transformative.

Positive technologies have been proposed to support positive psychology with artifacts to mediate purposeful actions with measurable effects toward happiness. In broad terms, positive technologies have addressed [15]:

- Mental health: promotion of well-being and happiness in vulnerable patients, patients presenting symptoms such as depression, eating disorders, and other observable behaviors, and patients requiring emotion regulation.
- Neuro-rehabilitation: support for the treatment of patients with partial sight, motor-cognitive, neural, or age-related disabilities.
- Empathy and pro-social behavior: support for patients with partial disabilities in the development of empathy, patients with learning disabilities, or patients under stress due to lack of cross-cultural social integration.
- Self-transcendence: promotion of experiences that generate a state of *awe*.

Positive technologies stand on three pillars:

1. Intervention: technology-mediated construction of experiences that can induce states related to hedonia and eudaimonia.
2. Monitoring: (possibly quantified) assessment of the effectiveness of intervention through measurable surrogate markers based on hemodynamic parameters [47], goosebumps [9, 36], monitoring and classification of micro-expressions, and posture and gesture patterns [11].
3. Assessment: interpretation of monitored values, often based on correlations between observations of markers and answers to standardized questionnaires such as Self-Regulation Questionnaires [8, 27], Emotion-Regulation Questionnaires [21], and Technology-Based Experience of Need Satisfaction Questionnaires [33].

Positive computing narrows positive technologies down to computer-based technologies, with particular emphasis on software-based technologies [7, 32].

Positive artificial intelligence has arisen as a recent proposition to characterize intelligent positive computing [42]. Following the general literature about artificial intelligence [38, 41], the field of artificial intelligence can be summarized as the scientific and technological development of building systems that can be organized along two attribute scales:

1. *Structural/Imitative intelligence:*

   - Systems that are *inherently* intelligent by featuring structural organization that aligns with explanatory theories of intelligence applicable to biological agents *versus*
   - Systems that are capable of *imitating* intelligent behavior, regardless of their organization; and

2. *Human/Mathematically defined rationality:*

- Systems that adopt as reference for intelligence/intelligent behavior human (or other biological) agents *versus*
- Systems that adopt as reference some theory of rationality that characterizes "optimal" intelligence in terms of efficiency in goal seeking.

Given its technological orientation, positive artificial intelligence is focused on *imitative intelligence*, grounded either on "human"/psychological or on mathematically defined theories of rationality, and is geared toward the development of user-centered intelligent agents. For these reasons, another dimension for analysis and design of intelligent agents must be considered, namely the *degree of awareness of social interactions* of intelligent agents, in order to characterize the extent to which a designed agent includes a consideration of relations with other agents—which can be other intelligent agents or human (or other biological) agents. For the sake of concision, the spectrum of possibilities of awareness of social interactions can be reduced to four values:

1. *Egocentric agents*, which only account for their own goals and resources and consider any other entity and event in the environment as either resources to be exploited or barriers to be overcome. Such agents correspond to what was considered during the pioneering developments of artificial intelligence.
2. *Strategic agents*, which are aware of the existence of other agents endowed with goals and resources, but still give full priority to the management of their own resources to reach their own goals. These agents assume that the other agents will behave similarly and build strategies that, in order to optimize their own goals and use of resources, may be mutually beneficial to other agents. Such agents correspond to what is considered in classical economics and mathematical game theory [41].
3. *Social agents*, which take into account collective goals and resources and act to optimize them, considering long-term goals that may outlive the agents themselves. Such agents consider the benefit of the society of agents as a whole and of future generations as well as their own.
4. *Empathic agents*, which are capable of "wearing other agents' shoes," balancing the importance of their own goals and those of other agents and deciding upon actions based on social emotions [22].

Each value on this spectrum adds to the previous one as a refinement and, as a consequence, increases the complexity of modeling agent interactions: Strategic agents are egocentric agents *plus* awareness of the existence of other agents; social agents are strategic agents *plus* awareness of collective goals; and empathic agents are social agents *plus* sensitivity motivations based on social emotions.

Recently, machine learning has become increasingly prominent in artificial intelligence, promoting the view that intelligent agents should update their behavior dynamically as new evidence (in the form of collected data) is gathered. Most frequently, agents are designed to operate in two separate modes:

1. *Training mode*: Evidence gathering is performed and the behavior of agents is updated accordingly in order to build a new version of each agent.

2. *Execution mode*: Agents are deployed for problem solving based on a stable set of actions obtained from the latest iteration in *training mode*.

Depending on how costly the *training mode* or the implementation of the *execution mode* can be, the *training–execution* cycle can be frequent, rare, or one-shot (meaning that *training* can occur only once and an intelligent agent can be deployed for execution for the rest of its life cycle). One-shot machine learning can be envisaged as the application of machine learning techniques to support the design of artifacts. *Positive design* is the sub-field of design theory devoted to the development of artifacts to support and foster human flourishing [12]. The conflation of *positive design, positive artificial intelligence,* and one-shot machine learning leads to the concept of *positively intelligent design agents*.

A *positively intelligent design agent—PIDA—*is an intelligent agent primarily built to support design activities focusing on the development of artifacts to promote positive experiences for human agents. In the following section we consider urban spaces as such artifacts and the development of *PIDAs* for urban design, so that evidence gathering can be used in a *training mode* to guide the design of urban landscapes, and *execution* can be envisaged as the ways in which constructed spaces enable and nudge their inhabitants toward health and well-being.

9.3 Affective Walkability and Urban Design

Ultimately, intelligent agents are designed and built to serve human needs, hence—directly or indirectly—intelligent agents exist to interact with humans. In some cases the interactions are evident, e.g., when agents actively interact with humans as artifacts designed for end users. In other cases the interactions can be subtle, e.g., when agents quietly provide humans with other artifacts that enable and inspire behaviors and emotional states. In all cases, consideration of the widest possible spectrum of consequences of interactions with intelligent agents is advised, and has become object of attention of scholars and regulatory agencies.

The ACM and the IEEE have prepared general codes of ethics for professionals in computer science and engineering, catering specifically for autonomous and intelligent agents in specialized sections [3, 43], and specialized academic research institutes and laboratories have been structured to work on how to ensure that intelligent agents are used to promote well-being, following carefully crafted ethics guidelines.[1]

Adopting a terminology suggested by Peters et al. [34], there has been an imbalance toward *nonmaleficence* over *beneficence* in the design of intelligent agents, i.e., codes and regulations have focused on what should be avoided to prevent harmful interactions, instead of what should be ensured to promote positive interactions (an important exception to this trend being the work of Peters et al. [33, 34]).

---

[1] See e.g., *Positive Computing* (http://www.positivecomputing.org/) and the *AI Now Institute* (https://ainowinstitute.org/).

A possible justification for this imbalance is the shorter time span required to assess *nonmaleficence* than *beneficence*. Ethical frameworks can help justify this statement. Ethical frameworks are commonly based on three *fundamentals* and four *principles* [5]:

- *Fundamentals:*

  1. Consequences and utilities of actions;
  2. Norms and duties;
  3. Virtues and capabilities.

- *Principles:*

  1. Nonmaleficence;
  2. Autonomy;
  3. Beneficence;
  4. Justice.

Nonmaleficence relates to ensuring that an intelligent agent does not cause or promote harm to humans. This can in many cases be verified based on a formal analysis of design and/or empirical testing. It also relates to consequential and utilitarian ethics, as it provides designers with information about negative effects that can amount to negative results (hopefully with low probability), which should be balanced with potential (and hopefully more likely) positive results. Autonomy relates to ensuring that an intelligent agent does not restrain humans from behaving according to their will. This can be more difficult to assess, as it relates to ensuring alignment with norms and duties and is also grounded on avoiding negative consequences of interactions. Beneficence, in contrast, relates to ensuring that an intelligent agent *does* promote the health and well-being of humans through interactions by supporting and promoting the flourishing of human capabilities. Finally, justice also relates to ensuring that humans can find a sense of fulfillment through social interactions based on a sense of promoting as well as living in a society based on fairness, trust, and justice. Intelligent agents can nurture justice and virtue *in* humans, even if as designed artifacts they cannot store such concepts per se.

The assessment of ethical behavior of intelligent agents requires more time and is harder to measure as we move from *negative* issues to be avoided toward *positive* issues to be promoted. The corresponding issues also move from *basic survival* through *hedonia* to *eudaimonia*. It seems natural, therefore, that *basic* and *negative* issues are easier to assess and to regulate. When dealing with artifacts that can be costly and time consuming to implement—such as urban spaces—this rationale fails, however, and a focus on virtue-based, positive, and eudaimonia-related attributes becomes relevant.

Upon conceiving urban spaces as artifacts whose design can be supported by intelligent agents that interact with humans with the purpose of promoting health and well-being, it seems natural to ask whether evidence could be gathered about existing interactions and their effects on citizens in such a way that certain patterns could be identified that could steer urban design toward improved interactions.

In order to answer this question, three types of resources are necessary:

1. A set of *potential interventions* in the form of coherent transformations that could be applied to urban spaces in order to optimize certain parameters, e.g., reorganization of sidewalks to induce walking patterns in pedestrians.
2. A set of *conceptual parameters* to be optimized that could act as indicators of relative distance toward a *goal state* characterized by the maximization of desired attributes, e.g., appreciation of architectural beauty and a sense of connection with places and communities.
3. A set of *objective and measurable surrogate attributes* that can be monitored and optimized and can be assumed to correlate well with the *conceptual parameters*, e.g., physiological markers whose dynamics of fluctuations indicate level of stress, states of awe, flow, etc.

With these resources, and once a dataset is obtained containing links between *interventions* and estimated *conceptual parameters* as characterized by values of *surrogate attributes*, the appropriate use of machine learning techniques can be employed to train a function that correlates *interventions* with *surrogate attributes* in such way that a candidate set of *interventions* can be identified to generate the optimal set of values of attributes corresponding to maximization of the *conceptual parameters*.

Particular attention can be given to walking activities, given that the majority of the daily life activities of citizens, e.g., related to physical conditioning and sports, consumer life, and social interactions, take place in a walking neighborhood [23]. In many urban settings, this is particularly relevant for vulnerable populations such as children, the elderly, and people with disabilities. Some studies indicate that physical activity plays an important role in population mental and physical healthcare [26]. A walking environment that is friendly is, therefore, a most relevant attribute for urban design [16].

The concept of *walkability* has been adopted recently as a conceptual parameter to indicate the capacity of a urban space to generate and sustain health and well-being [25, 28]. In summary, walkability measures the extent to which an urban setting is friendly for walking, considering aspects such as accessibility, comfort, safety, attractiveness in terms of architectural design and urban context, and efficiency with respect to the availability of services for pedestrians [18].

Affective walkability [4] introduces measurable attributes in the form of physiological markers that have been verified as effective in identifying affective states as means to assess walkability. Emotions often refer to mental states that arise spontaneously rather than consciously and are often accompanied by measurable physical and physiological fluctuations in human organs and tissues, such as the brain, heart, skin, blood flow, muscles, facial expressions, and voice. Being able to interpret the emotions of pedestrians in urban environments while interacting with other citizens, vehicles, and architectural and urban elements makes it possible to assess the extent to which an environment is perceived as safe, comfortable, and walkable. It can be argued—although it is a proposition that requires further empirical validation—that it is possible to also assess the extent to which an environment is capable of promoting a sense of belonging and reminding individuals of their views related to their purpose in life and achievements toward this purpose.

The assessment of affective walkability requires the design and performance of rigorous experiments to properly collect data. Thanks to the development of the technology, several sensors can be easily integrated into smartphones or wearable devices [48], making them more comfortable and usable even by elderly people. These signals could thus be valid indicators to assess quantitatively the safety perception of citizens when interacting with the surrounding environment.

9.4 Positively Intelligent Neighborhoods: A Call to Action

In the previous sections we reviewed the concept of positive artificial intelligence and introduced a perspective from which the design of urban landscapes can be supported by a peculiar category of positively intelligent agents coined *PIDA—positively intelligent design agents* capable of interacting with humans in subtle ways and in such a way that positive emotions and flourishing can be supported by the artifacts generated by *PIDAs*.

We have also presented how intelligent sensing technologies can be used to assess affective walkability, which in turn can be used to indicate the effectiveness of urban landscapes in the promotion of human flourishing. This way, given a dataset comprising measurements of physiological indicators capable of unfolding degrees of affective walkability, plus corresponding features of urban landscapes, we are capable in principle of building a *model* using machine learning techniques that can be used to:

1. Predict the degree of affective walkability that can be expected given features of a new urban landscape.
2. Predict the degree of affective walkability that can be expected from the same urban landscape given *simulated interventions* that can be performed in the urban scene.
3. Based on generative design techniques [2, 44], explore the space of potential interventions to find optimal interventions that can be put into effect.

In order to be able to carry out this research plan, we need to build a dataset as described here. In particular, we need to study how human affective states change due to space-varying conditions as well as to time-varying stimuli. To this end, two complementary types of experiments involving people of different age ranges and mobility capabilities should be performed:

1. **In-vitro** experiments, in which data collection is carried out in controlled laboratory environments. In this case, interference from external stimuli is limited and data quality is higher. The disadvantage of these experiments lies in the *unreal* nature of the stimuli.

Fig. 9.1 Wearable devices adopted for both in-vitro and in-vivo experiments

2. **In-vivo** experiments, in which data collection is performed in a real-life environment, which has the advantage of inducing more realistic affective state at the price of requiring the management of several sources of noise and, as a consequence, lowering data quality.

We propose for these experiments the integration of multi-modal signals from varied sources to define an affective walking assessment approach, considering data coming from physical activity and uncontrolled reactions related to affective responses to stressful conditions. Wearable sensors can be adopted to acquire photoplethysmography (PPG, which measures the blood volume registered just under the skin), which can be used to calculate the heart rate of the subject, and galvanic skin response (GSR, which measures the skin humidity due to sweating. Moreover, motion data, which can be physiological and captured by measuring the muscle activity with electromyography (EMG), as well as inertial data, should also be collected. The integration of appropriate transforms of these data sources has been empirically shown to be sufficient for studying pedestrian walkability. Several wearable devices are available on the market. We have first-hand experience collecting physiological as well as inertial data with the sensors Shimmer3 GSR+ and Shimmer3 EMG/ECG [6], shown in Fig. 9.1. Both sensors interface with ConsensysPRO software, also manufactured by Shimmer Co., which was used to setup trials, superimpose markers on raw data, and partially pre-process collected data.

With the adoption of these sensors, one experimental protocol for in-vitro experiments and two experimental protocols for in-vivo experiments are hereafter detailed in order to provide guidelines for other experiments and facilitate data collection by researchers interested in this topic.

9.4.1 In-vitro Experimental Protocol

In order to investigate the behavior of citizens and assess their affective states under different walking conditions, and in particular during a dynamic collision avoidance task, an experimental protocol in a controlled laboratory environment was devised and empirically tested, considering three within-subject conditions (free walk, forced speed walk, and collision avoidance). The laboratory environment is depicted in

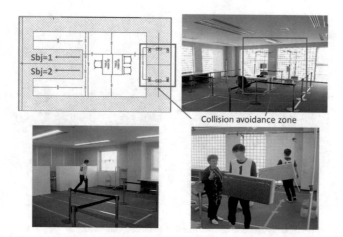

Fig. 9.2 In-vitro experiment. Top left: Plan of the laboratory showing the path chosen for the walking activities. The red rectangle identifies the collision avoidance zone in the two images of the first row. In this zone, two obstacles are moved by one of the experimenters, and the two subjects have to avoid collision (figure bottom right). During the rest of the path, subjects walk at their own natural pace

Fig. 9.2, The protocol of the experiment, which lasts about 30 min, can be summarized as follows:

1. **Free walk**: Two subjects at the same time walk freely without obstacles or speed constraints back and forth along the path depicted in Fig. 9.2.
2. **Collision avoidance**: Two subjects at the same time walk at their own pace along the path, one clockwise and the other counterclockwise. At about the halfway point of the path, they reach the collision avoidance zone, where they have to avoid collisions with both an obstacle (a swinging pendulum) and the other subject. They then complete the path at their natural pace and return in the opposite direction while repeating the same actions. This task thus consists of a free walking activity approaching and leaving the obstacle zone, and the effective collision-avoidance action.
3. **Forced speed walk**: Two subjects walk with a forced speed based on a metronome ticking along the same path. Three speeds are considered: **F1** $= 70$ bpm, **F2** $= 85$ bpm, and **F3** $=100$ bpm. At the end, a questionnaire is provided to the participants to gather information about the preferred walking frequency among these three.

Tasks should be separated by a period of rest, called **Baseline acquisition**, of about 1 min, to gather reference physiological responses. The whole procedure should be repeated three times.

This protocol has been used for data acquisition and analysis. More details and the results of data analysis of this experiment are reported in previous publications by certain authors of this chapter [4, 17].

Fig. 9.3 Examples of busy roads in the center of the Italian city of Cantu

9.4.2 In-vivo Experimental Protocols

Several factors contribute to changing a citizen's perception of safety and wellness while walking in the city, including age, sex, disability, and cognitive impairment. Moreover, other variables deriving from the environment should be considered, such as uneven pavement in sidewalks and crossings, the presence of loud noise, bulky work vehicles, poor visibility of ongoing cars (e.g., due to cars parked along the side of the road), and a lack of signs that facilitate road crossing [23]. In Fig. 9.3, an example urban environment in an Italian city is depicted.

The aim of experimental protocols designed for real uncontrolled urban environments is to assess the stress levels, safety, and confidence reported by citizens when dealing with pedestrian areas of the cities where they live in order to define interventions to increase walkability. To this end, two different scenarios in which elderly people or people with motor or cognitive deficits are particularly exposed can be considered to assess safety perception while walking:

1. Crossing a busy road without the aid of traffic lights, and
2. Walking on a long path within the city with alternation between comfortable and stressful paths.

9.4.2.1 Crossing a Busy Road Without the Aid of Traffic Lights

An ideal environment for this experiment is configured by:

1. A busy road with car congestion,
2. A lengthy crossing, and
3. The absence of pedestrian traffic lights.

Fig. 9.4 An example urban environment suitable for in-vivo experiments

Care should be taken to prevent risk for the subjects participating in the experiment. It is expected that the more dangerous the crossing, the greater the emotional arousal and the better the signals acquired. An example of such an environment is shown in Fig. 9.4, featuring a large number of vehicles, including work vehicles and trucks, especially at rush hour.

The experiment accounts for two different walking scenarios that are supposed to be related to different perceptions of safety: Free walking on sidewalks, and crossing a two-way road in correspondence to a crossroad without traffic lights. The sidewalk chosen has to be wide and comfortable. The chosen crossing, in contrast, can be considered moderately dangerous for the pedestrians. Besides the acquisition of physiological signals, as previously described, the experimental protocol also includes self-assessment questionnaires, both to evaluate the self-esteem levels of the participants and elicit the level of safety perception of each crossing.

The whole protocol is described as follows:

- Baseline: 2 min to acquire baseline physiological signals, when subjects stand straight and still to allow physiological responses in absence of any tasks to be recorded.
- Questionnaire completion: Rosenberg Self-Esteem Scale [37].
- Experiment Core (repeated 4 times):

  - Walking on sidewalk (non-stressful task).
  - One-minute baseline recording in which subjects stand straight and still to record physiological responses in absence of any tasks, also intended to bring subjects back to a *neutral* state before the next task.
  - Crossing the road and coming back to the start point (stressful task). In order to better understand subjects' behavior, this task can also be video recorded using a full HD camera. Every participant should complete an informed consent form to permit the recordings.
  - One-minute baseline, same as before.

Fig. 9.5 Experimental protocol for road crossing

   – Completion of crossing questionnaire.

• End of trial.

The ordering of the walking and crossing tasks should be randomly selected for each subject in order to avoid observation bias.

The whole experiment is planned to last for about 20 min, being long enough to gather usable data and short enough to prevent the subjects from becoming accustomed to the task at hand. The protocol is depicted in Fig. 9.5.

9.4.2.2 Walking Along a Long Path

The second experimental protocol lasts longer than the first and is devised to test stress levels over a longer time span. Subjects are asked to walk along a path that is initially comfortable, on even ground and, wherever possible, shaded and without road crossings or dangers of any kind. In the second part of the walk, subjects are asked to continue on uneven ground on a sidewalk near high-speed roads and where at least two difficult crossings are encountered. An example of a suitable pathway for this experiment is shown in Fig. 9.6. It begins at the entrance to a park with only pedestrian or bicycle-shaded areas, and continues in this way for about 300 meters. Halfway along the path, a crossing marks the beginning of the second part. The crossing itself is particularly difficult to get through, as it is quite wide and corresponds with a sub-urban road where cars pass at high speed. The pathway continues on a rough sidewalk next to a busy and dangerous road to the final checkpoint in the parking lot of a shopping center.

The experimental protocol is described as follows:

• Baseline: Two-minute session to acquire reference physiological signals, where subjects stand straight and still to record physiological responses in the absence of any tasks.
• Questionnaire filling: Rosenberg Self-Esteem Scale [37].
• One-minute baseline recording, where subjects stand straight and still to record physiological responses in the absence of any tasks.
• Walking in the park for about 300 m (non-stressful task).
• One-minute baseline recording, same as before, also intended to bring subjects back to a *neutral* state before the next task.

Fig. 9.6 A suitable pathway for the path-walking experiment

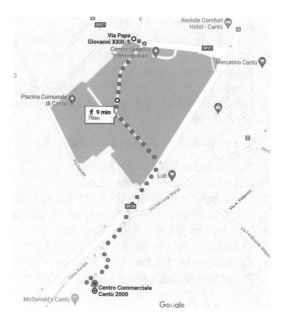

- Crossing the sub-road (stressful task).
- Walking along the sidewalk for about 300 m, next to a busy and stressful road, to a shopping center.
- One-minute baseline, same as before.
- Questionnaire completion.

The order of the two walking tasks should be randomly selected for each participant to avoid observation bias. The Rosenberg Self-Esteem Scale measures the self-appreciation and self-confidence. A Likert scale from 1 (*Absolutely not*) to 4 (*Absolutely yes*) is adopted and the items are:

1. I feel that I'm a person of worth, at least on an equal level with others.
2. I feel that I have a number of good qualities.
3. All in all, I am inclined to feel that I am a failure.
4. I am able to do things as well as most other people.
5. I feel I do not have much to be proud of.
6. I take a positive attitude toward myself.
7. On the whole, I am satisfied with myself.
8. I wish I could have more respect for myself.
9. I certainly feel useless at times.
10. At times I think I am good for nothing at all.

Two customized questionnaires have been defined to collect subjective safety perceptions, scored on a three-value scale: *NULL, LOW, HIGH*. The items of the crossing questionnaire are:

1. Stress level during the crossing.
2. Confidence level toward cars during the crossing.
3. Interference level by other means of transportation during the crossing.
4. Influence level from other pedestrians.
5. Confidence level in the crossing without traffic control or traffic lights.
6. Confidence level in the crossing with disturbing elements (parked cars, partially blocked view, etc.).

The items of the walking questionnaire are:

1. Stress level walking in the park.
2. Stress level while crossing.
3. Confidence level toward the cars during the crossing.
4. Interference level by other means of transportation during the crossing.
5. Influence level from other pedestrians.
6. Confidence level in the crossing without traffic control or traffic lights.
7. Confidence level in the crossing with disturbing elements.
8. Stress level walking along the sub-urban road.

9.5 Conclusion

In this chapter we show that through the adoption of intelligent sensing and positive technologies, positive artificial intelligence meets affective walkability, making it possible to develop the concept of positively intelligent neighborhoods. Hence, we close this chapter with a *call to action* of those interested in the optimization of urban settings toward the development of truly *human-centered cities* and *positively intelligent neighborhoods*, which should start with data collection worldwide in such a way that it can be structured and stored in an open access data repository. An initiative along these lines could make good use of protocols similar to those presented in the previous section to collect data that can be structured and prepared to be consumed by *PIDAs*, coupled with generative design techniques, which can then be used by urban designers and architects to design *positively intelligent neighborhoods*.

References

1. Alexopoulos, C., Pereira, G.V., Charalabidis, Y., et al.: A taxonomy of smart cities initiatives. In: Proceedings of the 12th International Conference on Theory and Practice of Electronic Governance, pp. 281–290 (2019)
2. van Ameijde, J., Song, Y.: Data-driven urban porosity-incorporating parameters of public space into a generative urban design process. In: Learning, Adapting and Prototyping, Proceedings of the 23rd International Conference of the Association for Computer-Aided Architectural Design Research in Asia (CAADRIA). CUMINCAD (2018)

3. Anderson, R.E.: ACM code of ethics and professional conduct. Commun. ACM **35**(5), 94–99 (1992)
4. Bandini, S., Gasparini, F.: Social and active inclusion of the elderly in the city through affective walkability. The Review of Socionetwork Strategies, pp. 1–17 (2021)
5. Bartneck, C., Lütge, C., Wagner, A., et al.: An Introduction to Ethics in Robotics and AI. Springer Nature (2021)
6. Burns, A., Doheny, E.P., Greene, B.R., et al.: Shimmer™: an extensible platform for physiological signal capture. In: 2010 Annual International Conference of the IEEE Engineering in Medicine and Biology, pp. 3759–3762. IEEE (2010)
7. Calvo, R.A., Peters, D.: Positive Computing: Technology for Wellbeing and Human Potential. MIT Press (2014)
8. Carey, K.B., Neal, D.J., Collins, S.E.: A psychometric analysis of the self-regulation questionnaire. Addict. Behav. **29**(2), 253–260 (2004)
9. Chirico, A., Ferrise, F., Cordella, L., et al.: Designing awe in virtual reality: an experimental study. Front. Psychol. **8**, 2351 (2018)
10. Csikszentmihalyi, M., Abuhamdeh, S., Nakamura, J., et al.: Flow (1990)
11. Ekman, P.: Emotions revealed. Bmj 328(Suppl S5) (2004)
12. Faust, J.: Positive design. J. Am. Soc. Inf. Sci. Technol. **60**(9), 1887–1894 (2009)
13. Frankl, V.E.: Man's Search for Meaning. Simon and Schuster (1985)
14. Froh, J.J.: The history of positive psychology: truth be told. NYS Psychol. **16**(3), 18–20 (2004)
15. Gaggioli, A., Villani, D., Serino, S., et al.: Positive technology: designing e-experiences for positive change. Front. Psychol. 10 (2019)
16. Gaglione, F., Cottrill, C., Gargiulo, C.: Urban services, pedestrian networks and behaviors to measure elderly accessibility. Transp. Res. part D: Transp. Environ. **90**(102), 687 (2021)
17. Gasparini, F., Grossi, A., Nishinari, K., et al.: Emg for walkability assessment: a comparison between elderly and young adults. In: AIxAS@ AI* IA, pp. 88–100 (2020)
18. Gasparini, F., Giltri, M., Bandini, S.: Safety perception and pedestrian dynamics: experimental results towards affective agents modeling. AI Commun. **34**(10), 1–15 (2021)
19. Gehl, J.: Cities for People. Island Press (2013)
20. Gillham, J.E., Seligman, M.E.: Footsteps on the road to a positive psychology. Behav. Res. Therapy **37**(1), S163 (1999)
21. Gullone, E., Taffe, J.: The emotion regulation questionnaire for children and adolescents (erqca): a psychometric evaluation. Psychol. Assess. **24**(2), 409 (2012)
22. Hareli, S., Parkinson, B.: What's social about social emotions? J. Theory Soc. Behav. **38**(2), 131–156 (2008)
23. Kim, H.: Wearable sensor data-driven walkability assessment for elderly people. Sustainability **12**(10), 4041 (2020)
24. Kitson, A., Prpa, M., Riecke, B.E.: Immersive interactive technologies for positive change: a scoping review and design considerations. Front. Psychol. **9**, 1354 (2018)
25. Le, T.P.L., Leung, A., Kavalchuk, I., et al.: Age-proofing a traffic saturated metropolis-evaluating the influences on walking behaviour in older adults in ho chi minh city. Travel Behav. Soc. **23**, 1–12 (2021)
26. Lee, G., Choi, B., Jebelli, H., et al.: Wearable biosensor and collective sensing–based approach for detecting older adults' environmental barriers. J. Comput. Civil Eng. **34**(2), 04020,002 (2020)
27. Levesque, C.S., Williams, G.C., Elliot, D., et al.: Validating the theoretical structure of the treatment self-regulation questionnaire (tsrq) across three different health behaviors. Health Educ. Res. **22**(5), 691–702 (2007)
28. Lo, R.H.: Walkability: what is it? J. Urban. **2**(2), 145–166 (2009)
29. Matan, A., Newman, P.: People cities: The life and legacy of Jan Gehl. Island Press (2016)
30. Montgomery, C.: Happy City: Transforming Our Lives Through Urban Design. Penguin UK (2013)
31. o'Brien, T.: Douglas Engelbart's Lasting Legacy. San Jose Mercury News (1999)

32. Pawlowski, J.M., Eimler, S.C., Jansen, M., et al.: Positive computing. Bus. Inf. Syst. Eng. **57**(6), 405–408 (2015)
33. Peters, D., Calvo, R.A., Ryan, R.M.: Designing for motivation, engagement and wellbeing in digital experience. Front. Psychol. **9**, 797 (2018)
34. Peters, D., Vold, K., Robinson, D., et al.: Responsible ai-two frameworks for ethical design practice. IEEE Trans. Technol. Soc. **1**(1), 34–47 (2020)
35. Peterson, C., Park, N., et al.: Meaning and positive psychology. Int. J. Existent. Posit. Psychol. **5**(1), 7 (2014)
36. Quesnel, D., Riecke, B.E.: Are you awed yet? how virtual reality gives us awe and goose bumps. Front. Psychol. **9**, 2158 (2018)
37. Rosenberg, M.: The association between self-esteem and anxiety. J. Psych. Res. (1962)
38. Russell, S.J., Norvig, P.: (2020) Artificial Intelligence: A Modern Approach (4th Edition). Pearson (2020)
39. Seligman, M.E.: Flourish: A Visionary New Understanding of Happiness and Well-Being. Simon and Schuster (2012)
40. Seligman, M.E., Csikszentmihalyi, M.: Positive psychology: an introduction. Am. Psychol. **55**(1), 5–14 (2000)
41. Shoham, Y., Leyton-Brown, K.: Multiagent Systems: Algorithmic, Game-Theoretic, and Logical Foundations. Cambridge University Press (2009)
42. Correa da Silva, F.S.: Towards positive artificial intelligence. In: International Conference of the Italian Association for Artificial Intelligence. Springer, pp. 359–371 (2020)
43. Spiekermann, S.: IEEE p7000-the first global standard process for addressing ethical concerns in system design. Multidiscip. Digit. Publ. Inst. Proc. **1**(3), 159 (2017)
44. Tang, Z., Ye, Y., Jiang, Z., et al.: A data-informed analytical approach to human-scale greenway planning: integrating multi-sourced urban data with machine learning algorithms. Urban For. Urban Greening **56**(126), 871 (2020)
45. Taylor, E.: Positive psychology and humanistic psychology: a reply to Seligman. J. Hum. Psychol. **41**(1), 13–29 (2001)
46. Winograd, T.: Shifting viewpoints: artificial intelligence and human-computer interaction. Artif. Intell. **170**, 1256–1258 (2006)
47. Yamaguchi, D., Tezuka, Y., Suzuki, N.: The differences between winners and losers in competition: the relation of cognitive and emotional aspects during a competition to hemodynamic responses. Adap. Hum. Behav. Physiol. **5**(1), 31–47 (2019)
48. Yetisen, A.K., Martinez-Hurtado, J.L., Ünal, B., et al.: Wearables in medicine. Adv. Mater. **30**(33), 1706,910 (2018)

Chapter 10
Machine Tells Us New Potential Values-Physics, Perception, and Affective Evaluations

Maki Sakamoto and Yuji Nozaki

Abstract This chapter introduces an AI system that understands the multidimensional meanings of Japanese sound-symbolic words expressing a texture sensation. Our experiments have shown the systematic association between linguistic sounds and textures in Japanese sound-symbolic words. In Japanese, it is possible to produce almost an infinite number of sound-symbolic words by combining sounds associated with certain perceptual impressions, and any kind of texture can be expressed by combining linguistic sounds. The AI system converts the meaning of a sound-symbolic word into the information equivalent to evaluations against 26 pairs of texture adjectives based on an analysis of the sounds of the word. We also developed another AI system that generates a new sound-symbolic word based on the association between Japanese linguistic sounds and subjective impressions, namely, the perceptual feeling of the texture of various materials/objects. This method goes beyond AI in handling existing linguistic data by supporting the development of new values and materials through the interaction of people with new sound-symbolic words. In general, when people notice a new texture that cannot be expressed by existing words, they create words to express that texture. If a machine notices a new texture and can generate a texture expression following the system introduced in this chapter, this can make people aware of the potential value of things, new products will be born and the physical world will be changed.

M. Sakamoto (✉) · Y. Nozaki
University of Electro-Communications, Chofu, Japan
e-mail: maki.sakamoto@uec.ac.jp

© Springer Nature Switzerland AG 2023
Y. Ohsawa (ed.), *Living Beyond Data*, Intelligent Systems Reference Library 230,
https://doi.org/10.1007/978-3-031-11593-6_10

10.1 What Is This? Simple and Typical Data for Machine Learning

When people see the many objects around them, they recognize them instantaneously and involuntarily; there is no 'need to wait for a few minutes after looking at a chair to grasp that it is in fact a chair. Machines have great difficulty in accomplishing this task, and researchers in the field of computer vision have been working for decades to find a solution to this problem. We know the mechanism by which visual data enter the human visual system and how they are processed, but we are still unsure how our brain categorizes and organizes the data. Researchers have extracted features from images and had machines learn from them. These images show variations in factors like size, angle, perspective, occlusion, and illumination, so the same object looks very different to a machine when it is presented from a different perspective. Humans, on the other hand, will immediately recognize an object from any angle. Even if a chair is partially blocked, we can still identify it. Machines will fail in this situation because each view is essentially a new object for them. In the last decade, several features and algorithms have been formulated to address this problem. These features are invariant to scale, rotation, and to some extent illumination. Examples include SIFT, SURF, PHOG, FAST, BRIEF, and Pyramid Match Kernel. Each feature descriptor has its own advantages and disadvantages. Researchers have tried to develop formulations that are close to those of the human visual system so that recognition accuracy increases. Object identification and object recognition are part of the research area of computer vision in the field of artificial intelligence (AI), which focuses on training machines such as robots to recognize different objects. As AI machines increasingly become part of our everyday lives, machine learning is expected to make improvements in object identification skills. Computer vision is the science of computer systems recognizing and analyzing different images and scenes; a key component of computer vision is thus object identification and recognition, which is used to perform a number of AI tasks such as facial recognition, vehicle detection, and self-driving. The object recognition process is based on CNNs, namely convolutional neural networks, which are designed to focus on neighboring pixels within images. Each image is passed through the network as an input and then sent back as an output with each object classified. Many researchers have sought to improve object recognition accuracy and speed, which is a task critical to the future of AI, for the better that robots can understand their surroundings, the better able they are to perform complex tasks. As a result, a number of recent advancements in deep learning have solved object recognition problems, as deep neural networks have shown a high level of performance on object classification tasks, which is especially useful for surmounting the challenge facing AI systems of employing the accurate detection of fast-moving objects like vehicles. Typical computer vision methodology has emphasized the role of labeled data. For example, to build a face detector, one needs a large collection of images labeled as containing faces. The need for large labeled data sets poses a significant challenge for problems where labeled data are rare. Therefore, Quoc V. Le et al. investigated the feasibility of building high-level features from only unlabeled data [1], which would provide an inexpensive way to develop features from unlabeled data. In this way, machines can now correctly recognize objects as well as or better than humans. Like humans, machines are expected to correctly recognize a cat as a cat and a chair as a chair. However, that is a story about data that has a correct answer. "A cat is a cat" is the correct answer, and "a chair is a chair" is the correct answer.

10.2 What's This Texture? Affective Data Could be Infinite

10.2.1 Texture Expression

The answer to the question "What's this?" is simple: "It is a cat." The question has a correct answer. There is a simple symbolic relationship between the object/concept and the name/word. On the other hand, the answer to the question "What's this texture?" or "How does this feel to you?" could be infinite and vary among people. The question has no correct answer. Despite recent advances in material recognition, high-level human cognition such as texture recognition remains one of the most challenging open problems. It is still too difficult for computer vision to express texture like humans because texture recognition has no correct or incorrect answer; it is ambiguous. Things in the world have their own unique textures. Texture recognition is achieved by processing in which various physical quantities (surface shape, color, etc.) of objects are detected by human sensory receptors and perceived in the brain. Material and texture perception has been studied in various fields such as neuroscience, psycho-physics, and vision psychology and has been revealed to involve glossiness, transparency, wetness, and roughness perceptions [2–9]. Although humans perceive texture almost unconsciously and express it easily, there is no computer that can express the textures of materials using texture terms. Moreover, the texture of the products or materials we encounter in daily life is not represented by a single texture-related adjective. That is, a product or material is usually represented by two or more texture-related adjectives. For example, the texture of a down quilt can be expressed as softness and a light and fluffy texture and the texture of sandpaper can be expressed as a dry and rough sensation. In this article, we focus on a word class that can express textures with one word.

The relationships between speech sounds and word meanings are arbitrary in general, such as /dog/ as a dog or /Hund/ as a dog in German. On the other hand, recent studies have suggested that iconicity (aspects of form resembling aspects of meaning) rather plays some role in structuring vocabulary (for a review, see [10]). Some studies have studied phonological iconicity, i.e., "sound symbolism," as a help to understand the development of language abilities (e.g., [11–13]) and language evolution ([14, 15]). Over the decades many researchers have demonstrated the existence of sound symbolism (e.g., [16–18], for early studies) in languages worldwide (e.g., [19–29]). Even in Indo-European languages such as English, there is clear sound symbolism ([30–32]). For example, roughly half of the English words starting with "gl-" imply something visually bright, as in "glance," "glare," "gleam," "glimmer," "glamour," "glass," "glaze," "glimpse," "glint," "glisten," "glitter," "globe," "glossy," and "glow." Landmark studies have found sound symbolism in words referring to visual shapes such as "mal" verses "mil" and "bouba" verses "kiki" for round verses sharp shapes, respectively ([18] and [15], respectively). Sound symbolism in vision has been discussed in terms of the process by which speakers link phonetic features with meanings [33]. For instance, speakers associate nonsense words articulated with rounded lips, such as "bouba," with round shapes, and nonsense words articulated without rounded lips, such as "kiki," with sharp shapes. In addition, the numbers of studies investigating sound symbolism in gustation are growing (e.g., [34, 35]). Gallace, Boschin, and Spence (2011), for example, showed that salt and vinegar crisps such as potato chips tend to be rated more often as "kiki" than cheddar cheese, yoghurt, or jam, and chocolate with mint chips and crisps also tend to be more associated with "kiki" than regular chocolate.

Sakamoto and Watanabe explored sound symbolism in tactile texture [36], especially the association between the phonemes of Japanese sound-symbolic words (hereafter, SSWs) expressing tactile sensations of materials. Compared with other languages, Japanese has a large vocabulary of SSWs as texture words, and all Japanese phonemes are systematically associated with sound-symbolic meanings [37]. Japanese can express any kind of texture by combining linguistic sounds. One may say the texture of this cat is "fuwa-fuwa" and that of another cat is "howa-howa." Most studies of tactile perception have used semantic differential scales comprising multiple adjective pairs [38] to explore the fundamental dimensions of tactile perceptual space (e.g., rough/smooth, hard/soft, slippery/sticky, and warm/cold). Unlike previous studies, we have investigated the sound symbolism of texture SSWs, since SSWs classify perceptual space with finer resolution than adjectives can and the sounds of Japanese SSWs are closely related to perceptual qualities. In fact, an experiment examined the variety of words of the two types of expressions (SSWs or adjectives) when participants touched 40 materials and expressed their sensations using SSWs or adjectives [39]. In all, 279 types of SSWs in 1,200 trials (40 materials x 30 participants) were obtained versus 124 types of adjective words (less than half the number of SSWs). Furthermore, most Japanese SSWs expressing tactile sensations are formed with two-syllable reduplication (C1V1C2V2-C1V1C2V2, where C and V indicate consonant and vowel, respectively, e.g., "sara-sara"), and the sounds of the first syllable (C1V1, e.g., "sa" for "sara-sara") are strongly associated with evaluations of tactile sensations [37]. For instance, "sara-sara" and "zara-zara," which differ only in the first syllable of the reduplicated unit, denote totally different tactile sensations: While the former is used for expressing a smooth and pleasant touch, the latter is used to express a rough and unpleasant touch. In contrast, "beta-beta" and "beto-beto," though they only differ in the second syllable of the reduplicated unit, express almost the same tactile sensation, namely stickiness. Therefore, we expected that the first syllable of Japanese SSWs could be a guide for systematically exploring touch-sound correspondence.

10.2.2   Experiment Showing the Connection Between Texture and Words

We first examined the relationship between the first syllables of Japanese SSWs and tactile sensations that subjects experienced from materials. Various kinds of materials of different textures were prepared for this experiment, including glass, papers, stones, sand, metals, rubbers, woods, clay, and plastics. The tactile sensations of these materials were evaluated by the SD method.

A total of 15 native Japanese speakers aged between 19 and 26 years (10 male and 5 female) with no abnormalities in tactile sense participated as volunteer participants for the experiment. The participants were not told the purpose of the experiment in advance.

In all, 120 samples of different kinds were cut to sizes of $60 \times 60 \times 2$ mm. To eliminate visual dependency, samples were presented to a subject in a box with a square hole of 80×100 mm. To evaluate the tactile sensation of a material, all subjects used their index finger of their dominant hand. After touching the material, they described their tactile perceptions of the material with SSWs. As the general population lacks detailed knowledge of the usage of SSW, brief instructions on SSW were provided to all participants with examples (e.g., the difference between SSWs and non-SSW). To avoid biases, we chose non-tactile-related SSWs for the examples used in these instructions.

Our experiment was conducted in two steps, a describing step and a rating step. In the describing step, a subject touches a sample in a box freely with their index finger for up to 30 s and expresses their tactile perception with an SSW elicited by touching the material. In the rating step, the participant evaluates the sensations of the material with eight adjective pairs (comfort-discomfort, bumpy-flat, rough-smooth, hard-soft, non-elastic-elastic, slippery-sticky, dry-moist, and warm-cold) on scale of one to seven while touching the same material. We chose these eight pairs from a review paper of research on tactile perceptual dimensions [38]. The selected adjectives referred to comfort/discomfort and certain fundamental dimensions frequently used in previous tactile studies. No time limit was imposed for the rating sequence. Throughout the experiment, subjects were told that they could use any SSW, regardless of whether it was common or novel, to express their tactile perception. In Japanese, a novel SSW can be easily produced by combining parts of existing vocabulary (for example, "mofu-mofu," a newly produced sound-symbolic expression, is a combination of "moko-moko" and "fuwa-fuwa"). By using the spontaneously produced expression, one can produce novel expressions that were not included in the predefined testing vocabulary of the research design. While participants described and rated the tactile sensations, they were free to run their fingers along and push the surface of the materials to explore the samples' various properties. Each participant performed 120 trials (a single trial for each sample). The order of sample presentations was different for each subject.

As a result, we obtained 1,800 combinations of sound-symbolic expressions and sensory responses (120 samples $\times$ 15 participants) for each adjective pair. In 87.1%

of all trials (1,566 cases), the sound-symbolic expressions were of the two-syllable reduplicated form. Considering that the Japanese linguistic form itself also has a sound-symbolic meaning [37], we used the first syllables of these 1,566 instances to examine sound symbolism in touch. Examining the results for all answers, we found that novel sound-symbolic expressions were used in approximately 20% of the cases. We also found the 22 first syllables that occurred more than 16 times (1% of 1,566) and used these 22 syllables to explore sound symbolism in touch.

Table 10.1 shows a summary of the evaluation result of 22 syllables for each adjective pair. Values that found to be significant difference from the average of 1,566 cases (t-test, $p < 0.05$), and those the absolute values of whose divergence from the average were larger than 0.5 are shown in the table. For all adjective pairs, we found several types of associations between the first syllables and tactile qualities.

Figure 10.1 shows the result of hierarchical cluster analysis (Ward's clustering algorithm) performed on the mean values of the 22 syllables in terms of eight adjective pairs. The first set of sequences (/ts/ + /u/ and /s/ + /u/) is related to comfortable, flat, smooth, hard, non-elastic, slippery, dry, and cold evaluations (see also Table 10.1). These phonemes were obtained when touching materials such as glass, and metals. The second set of sequences (/s/ + /a/, /p/ + /a/, /p/ + /o/ and /k/ + /a/) was related to the slippery and dry evaluations obtained for materials such as paper, and plastics. The third set of sequences (/ɸ/ + /u/ and /m/ + /o/) was related to comfortable, bumpy, rough, soft, elastic, and warm evaluations, obtained from materials such as soft fabrics. The fourth set of sequences (/g/ + /a/, /dʒ/ + /a/, /b/ + /o/, /dz/ + /a/, /g/ + /o/, and /t ç/ + /i/) was related to the discomfort, bumpy, rough, hard, non-elastic, and dry evaluations, obtained by materials such as woods, stones, and sandpapers. The fifth set of sequences (/b/+/u/, /g/+/u/, /p/+/u/, /ç/+/i/ and /p/+/i/) was related to the elastic, and wet evaluations, obtained from materials such as rubber. The sixth set of sequences (/b/+/e/, /n/+/e/, and /p/+/e/) was related to the uncomfortable, flat, smooth, soft, elastic, sticky, and wet evaluations obtained from materials such as slime.

In this way, our analysis showed the systematic associations between sound and tactile textures in Japanese SSWs. In Japanese, it is possible to produce an almost infinite number of SSWs by combining sounds associated with certain perceptual impressions, and any kind of texture can be expressed by combining linguistic sounds. Therefore, the answer to the question "What's this texture?" or "How do you feel about this?" could be infinite and vary across individuals. The question has no single correct answer. In the next section we shall consider how a machine deals with these affective data.

Table 10.1 Sound-symbolic associations of tactile perceptions

N = 1397/1566		Comfort-Discomfort (0.11)	Bumpy-Flat (−0.12)	Rough-Smooth (−0.06)	Hard-Soft (−0.18)	Nonelastic-Elastic (0.05)	Slippery-Sticky (0.25)	Dry-Moist (0.43)	Warm-Cold (−0.41)
/ts/+/u/	149	0.92	−1.87	−1.70	1.34	1.21	1.89	1.01	−1.18
/s/+/u/	94	1.13	−1.67	−1.78	–	0.87	1.70	–	–
/s/+/a/	130	0.68	–	−0.68	–	–	1.14	1.08	–
/p/+/a/	22	–	–	–	–	–	–	–	–
/p/+/o/	29	–	–	–	–	–	–	–	–
/k/+/a/	52	–	–	–	1.23	0.79	0.98	1.77	–
/ɸ/+/u/	118	1.30	–	–	−1.73	−1.16	–	–	0.91
/m/+/o/	32	1.19	1.06	1.00	–	–	–	–	1.00
/g/+/a/	50	–	1.40	1.28	0.84	–	–	1.00	–
/dʒ/+/a/	24	–	1.38	1.38	0.96	–	–	1.33	–
/b/+/o/	63	−0.56	1.95	1.27	0.87	0.78	–	–	–
/dz/+/a/	195	−0.49	0.87	1.52	0.48	0.88	–	1.24	–
/g/+/o/	52	–	1.33	0.92	1.27	1.00	–	1.10	–
/tɕ/+/i/	23	−0.87	1.35	2.52	–	–	–	1.43	–
/b/+/u/	27	–	–	–	–	−1.33	−0.74	−1.00	–
/g/+/u/	32	–	0.63	–	−1.44	−1.44	−0.66	−1.06	–
/p/+/u/	57	0.63	–	−0.98	−2.05	−2.21	–	−1.49	–
/ɕ/+/i/	18	–	–	–	–	–	–	−0.89	–
/p/+/i/	22	–	−1.00	–	–	–	–	−1.91	−1.82
/b/+/e/	101	−1.41	−0.96	–	−1.50	−1.00	−2.29	−1.35	–
/n/+/e/	26	−1.19	–	−0.88	−1.73	−1.04	−2.58	−1.81	–
/p/+/e/	81	−0.41	−1.23	−0.64	−1.00	−0.65	−1.49	−1.00	–

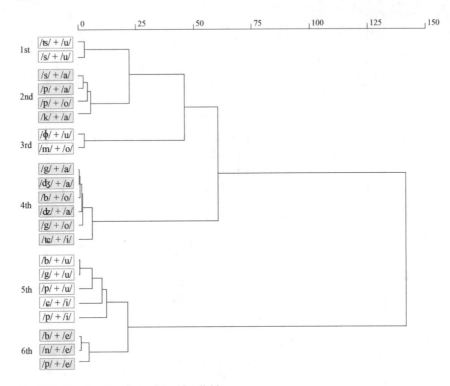

Fig. 10.1 Results of analysis of the 22 syllables

10.3 Machine Understands Texture Expressed by Sound-Symbolic Words

The sound-symbolic association between SSWs and texture suggested the idea of building a machine that understands the multidimensional meanings of a word by integrating the impression of each phoneme. In this section, referring to [40], we introduce a linear regression model that converts a Japanese SSW into quantitative ratings along multiple texture dimensions (26 pairs of adjectives). With this system, by analyzing the sounds of the word, for a given word that intuitively expresses a texture sensation, it is possible to obtain a predicted value of sensory evaluations on the 26 pairs of texture-related adjectives. An excellent advantage of this system is that a tactile sensation with many adjectival aspects can be determined from only a single word by analyzing the combinations of phonemes in the given word (even in newly created words) in Japanese.

To estimate the quantitative information of all possible SSWs, we built a database of sound-symbolic associations for all Japanese phonemes with the 26 pairs of adjectives based on the psychological experiments with the 78 paid participants. Then, to make a mathematical model of sound-symbolic associations between Japanese

phonemes and the scores of the 26 pairs of adjectives, we made a list of all possible combinations (14,584 combinations in total) of syllables in two-syllable forms
(i.e., /sara/) by combining all sounds in the Japanese. We also made a list of disyl-
labic reduplicated expressions (i.e., /sara-sara/), as this form frequently appears in
Japanese SSWs. In addition, we made another list of expressions by adding words
with all types of special phonemes used in Japanese SSWs (syllabic nasals /N/, gem-
ination /Q/, long vowels /R/, and adverbs ending in /Li/) to the two-syllable forms
(i.e., /saraLi/). Then, from these large lists, 312 words thought to be appropriate
to express texture perceptions were selected by three experts (including one of the
authors) from the perspectives of psychology and linguistics.

Participants were asked to respond how they felt by listening to the sound-
symbolic word by the SD method on a scale of 1–7 (e.g., for the comfortable-
uncomfortable scale, participants selected one of the following seven points: 1, *very
comfortable*; 2, *comfortable*; 3, *slightly comfortable*; 4, *neither*; 5, *slightly uncomfortable*; 6, *uncomfortable*; and 7, *very uncomfortable*). Answers were made by pushing
one of seven buttons placed in front of the monitor. Though participants were told
that there were no time limits for each trial, almost all trials took less than 1 minute.
The order in which SSWs were presented was randomized among participants, and
the polarities of the bipolar scales were also randomized in the answer matrix. As
a number of trials was huge (26 rating scales × 312 words = 8,112 trials), the 78
participants were divided into 6 groups of 13 participants. Each participant rated 52
words on 26 scales.

As the result of analysis, the average standard deviation of the ratings for each
scale was calculated to be 1.3, and 98% of the data was within 2.0 SD from the
average. To estimate the sensory impressions of SSWs, we built a model based on
Eq. (10.1) for predicting each rating value:

$$Y = \sum_{i=1}^{13} X_i + Const, \qquad (10.1)$$

where Y represents the rating value on the respective scale, and $X_1 - X_{13}$ correspond to the values defined in Table 10.2. $X_1 - X_6$, respectively, are the mean values
of consonant, voiced sound/p-sound, contracted sounds, vowels, semivowels, and
special phonemes in the first syllable; $X_7 - X_{12}$ are respectively the same categories
for the second syllable; and X_{13} denotes the presence or absence of reduplication
in the word. Using the averages of the rating values as the objective variables and
the variation of phonemes as the predictor variables, we applied mathematical quantification theory class I, which is a type of multiple regression analysis. Table 10.2
shows examples of the results of the analysis for each scale.

With Eq. (10.1), the rating values of the SSW can be calculated by the linear sum
of values $(X_1 - X_{13})$. Taking "sara" as an example, this SSW is composed of the first
mora /sa/ (/s/ + /a/) and the second mora /ra/ (/r/ + /a/). Therefore, the value of the
"rough-smooth" scale on a seven-point scale (*smooth* = 1 to *rough* = 7) $Y_{roughness}$
is estimated to be:

Table 10.2 Correspondences between variables and phonemes

Firstsyllable	Secondsyllable	Phonological characteristics	Variation of phonemes
X_1	X_7	Consonants	/k/, /s/, /t/, /n/, /h/,/m/, /y/, /r/, /w/ or absence
X_2	X_8	Voiced sounds/p-sounds	Presence (/g/, /z/, /d/, /b/, /p/) or absence
X_3	X_9	Contracted sounds	Presence (/ky/, /sy/, /ty/, /ny/, /hy/,/my/, /ry/, /gy/, /zy/, /by/, /py/) or absence
X_4	X_{10}	Vowels	/a/, /i/, /u/, /e/, /o/
X_5	X_{11}	Semi-vowels	/a/, /i/, /u/, /e/, /o/ or absence
X_6	X_{12}	Special sounds	/N/, /Q/, /R/, /Li/ or absence
X_{13}		Reduplication	Presence (ex. *sara-sara*) or absence

$$Y_{roughness} = /s/(X_1) + absence(X_2) + absence(X_3)$$
$$+ /a/(X_4) + absence(X_5) + absence(X_6)$$
$$+ /r/(X_7) + absence(X_8) + absence(X_9)$$
$$+ /a/(X_{10}) + absence(X_{11}) + absence(X_{12})$$
$$+ absence(X_{13}) + Const.$$

$$= (-0.05) + (-0.32) + (-0.05)$$
$$+ (0.46) + (-0.02) + (-0.03)$$
$$+ (-0.14) + (-0.1) + (0.05)$$
$$+ (-2.19) + (0.2) + (-0.02)$$
$$+ (0.05) + (0.01) + (3.75)$$

$$= 2.05$$

The estimated value of 2.05 suggests that "sara" is associated with a smooth impression, rather than rough. The multiple correlation coefficients R between the predicted values and the mean values of the actual ratings (the values obtained from the participants) were used as indicators of prediction accuracy. For 20 scales, values of R were between 0.80 and 0.90, and for the other 6 scales, they exceeded 0.90. This means that the proposed model was sufficient to estimate the impressions of SSWs by analyzing the phonemes and forms of the words.

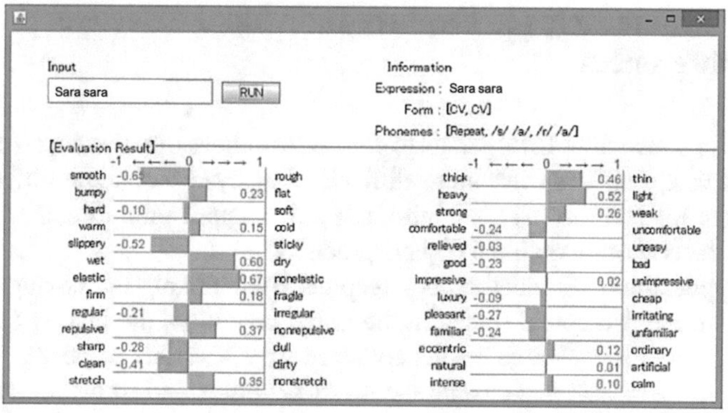

Fig. 10.2 Evaluation results for "sara-sara"

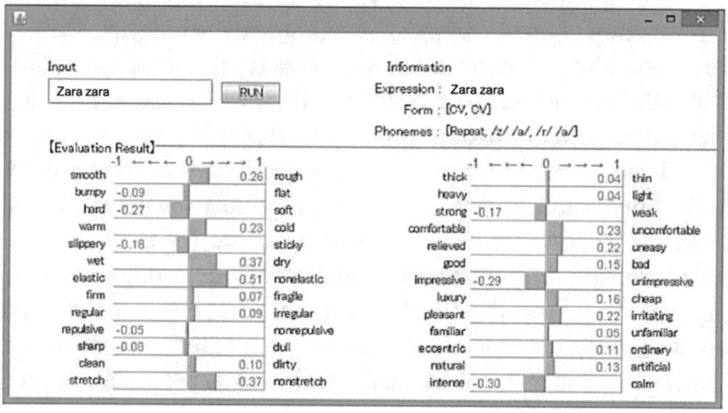

Fig. 10.3 Evaluation results for "zara-zara"

Figures 10.2 and 10.3 show the results of impression estimation for "sara-sara" and "zara-zara" done by our estimator. An SSW for which the user wants to convey its impression is inputted at the text field in the upper left. Then, triggered by "Run" bottom next of the text field, the system decodes the given SSW into a sequence of phonemes and classifies its form (e.g., presence of reduplicated structure). The analyzing module calculates the rating values of the word for the 26 scales and converts the calculated values to fall between −1 and 1. Finally, graphs of the estimated values of the word are displayed in the lower frame. The form and phonemic elements of the word are displayed in the upper-right frame.

10.4 Machine Helps to Capture Individual Affective Differences

The usage of SSW varies from person to person. To collect affective data is considered a difficult task, but it is even more difficult to analyze individual differences. In this section, referring to [41], we introduce a new application to collect how each consumer feels about individual objects/products.

This application individualizes a perceptual space simply by moving the SSWs projected on the perceptual space. In the initial condition, the map is generalized, being made from the average results collected from a variety of people. Users can intuitively move the SSWs with reference to the materials, and arrangements of SSWs as a whole in the perceptual space can be modulated according to an algorithm. Since the perceptual space map is an image, it is easily applied to machine learning.

To realize this application, we first listed up the most frequently used SSWs expressing perception of tactile sense by searching on the Internet. The 43 most used SSWs were selected for use in the initial word-based perceptual space. We then input each word into the system introduced in Sect. 10.3, which can convert an SSW into a multidimensional rating of texture properties to obtain a rating score from -1 to $+1$. With the database we introduced in the earlier section, which allows the user to estimate the impressions of a word by analyzing only the phonemes of the word, we performed PCA (principal component analysis) on the rating scores of the six fundamental dimensions of texture properties: "hard-soft," "rough-smooth," "bumpy-flat," "sticky-slippery," "wet-dry," and "warm-cold." The first and second principal components were used to visualize the initial word-based perceptual space. In the PCA plot, the distance between items in the word-based perceptual space represents the similarity in the multidimensional rating.

To guide users to move the SSWs in the perceptual space, we used 50 tactile materials [39] as references. With respect to the placement of reference materials, participants (six male and four female; mean age 22.8 years) evaluated the tactile impressions of the materials using the semantic differential method. For all participants, we used the same six adjective pairs as in the initial perceptual space: "hard-soft," "rough-smooth," "bumpy-flat," "sticky-slippery," "wet-dry," and "warm-cold." The participants rated their impressions of each material with 7-point scale (e.g., $-3 = $ *very smooth*, $3 = $ *very rough*). We then normalized the values obtained for each adjective pair with a mean of 0 and variance of 1. Placing 50 materials in the word-based perceptual space spanned by the first and the second principal components, we obtained a map with 43 SSWs (blue square and number) and 50 materials (red-circled number), as shown in Fig. 10.4.

In the application, users can modify the locations of SSWs freely as they indicate. They designate the placement of the SSWs on the perceptual space after considering the perceptual relationship between the materials and the SSWs. As the result, the application represents the individual's perceptual space as a two-dimensional map.

The system by Sakamoto and Watanabe (2019) is based on the algorithm described below. When a word "A" is moved, another word "B" is also moved according to

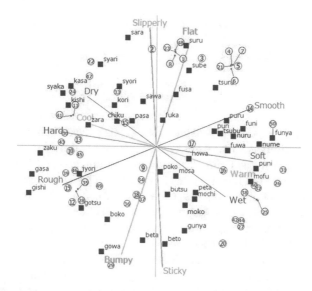

Fig. 10.4 Distribution map of the 43 SSWs and the 50 standardized tactile materials. Blue words indicate SSWs and red numbers indicate tactile materials. Six tactile scales, "hard-soft,", "rough-smooth," "bumpy-flat," "sticky-slippery," "wet-dry," and "warm-cold" are also shown. Reference [45] https://www.frontiersin.org/files/Articles/449345/fpsyg-10-01108-HTML/image_m/ fpsyg-10-01108-g001.jpg

the distance between the two words. The following equation is used to calculate the influence of moving A on the location of B.

$$\alpha = exp\left(\frac{-(d)^2}{2(\sigma)^2}\right). \tag{10.2}$$

Specifically, when word A is moved, the influence coefficient α (having a value of 0–1) exerted on word B is defined by formula (10.2), and word B is also shifted by α times the amount that word A moves. The influence coefficient is a Gaussian function, and as the distance (d: pixels) between words A and B increases, the coefficient decreases exponentially. The value of σ in the formula is a constant that defines the range of the influence of word A, and we define the value of σ as 95 in this algorithm. For example, if words A and B were separated by 95 pixels on an 800 × 600 pixel screen, the value of α would be 0.60 and word B would move 60 percent of the distance word A moves in the same direction as word A. If the separation were 190 pixels, the value of α would be 0.14, which means word B would shift by approximately 15%, whereas if the separation were 285 pixels, the value of α would be 0.01, which would have little effect on the movement of word B. In this application, when the user moves an SSW, its location can be fixed by clicking the right mouse button. When moving another word, the influence of fixed words needs to be considered. To calculate the influence of fixed word C, we use the following equation.

Similar to Eq. (10.2), this equation is based on a Gaussian function.

$$\beta = exp(\frac{-(d_1)^2}{2(\sigma)^2}) - exp(\frac{-2(d_2)^2}{2(\sigma)^2}). \qquad (10.3)$$

The influence coefficient β is defined by Eq. (10.3). For example, when word A is moved but word B is fixed, word C is affected by both A and B. The distance (d_1: pixel) between A and C (the first term) has an influence, as does the distance (d_2: pixel) between fixed B and C (the second term). The influence of word A is set to be larger than that of word B (the value of the numerator in the second term is $2d_2$). For example, in the case where words A, B, and C are sequentially arranged 95 pixels apart on the screen, when word A is moved, it influences word B by 0.6 and word C influences it by 0.14. Therefore, word C will move 0.46 times in the same direction as the movement of word A. If the distance between words B and C is equal to half the distance between words A and C ($d_2 = d_1/2$), the influence coefficient β will be 0 and word C will not move even if word A moves. When the user fixes the locations of words, the influence of these fixed words is added to the second term of β. Through these procedures, users change the SSWs in the perceptual space to match

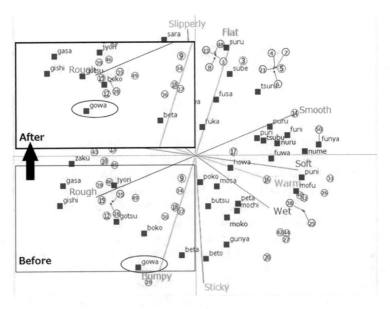

Fig. 10.5 Consequences of a change in the distribution map. In this figure, "gowa-gowa" has been moved to the upper left. Other SSWs, such as "gotsu-gotsu," "boko-boko," and "beta-beta," have been moved according to the influences calculated by Eqs. (10.2) and (10.3). Reference [45] https://www.frontiersin.org/files/Articles/449345/fpsyg-10-01108-HTML/image_m/fpsyg-10-01108-g002.jpg

their impressions. Figure 10.5 shows a state when a certain word is moved in the perceptual space. In this figure, "gowa-gowa" shown in the circle has been moved to the upper-left quadrant. The other SSWs, such as "gotsu-gotsu," "boko-boko," and "beta-beta," are then moved according to the influence coefficients calculated by Eqs. (10.2) and (10.3). The difference in each individual's perceptual space can be seen by comparing the distribution of SSWs. The number of movements may differ depending on the match between the initial perceptual space and the individual's perceptual space.

10.5 Machine Generates New Sound-Symbolic Words

As mentioned in the earlier chapter, humans have a very high-level command of SSWs. We even can coin a novel SSW that has never used by others to give a new affective impression to the listener. With this idea, machines affording us the ability to generate novel sounds may provide us a chance of becoming aware of new textures or undiscovered values. In this chapter, we will describe an algorithm to generate Japanese SSWs and the results of performance evaluations of models. We designed an algorithm that can generate SSWs with the affective impressions that a user desires by combining their phonemes and morphologies very flexibly.

10.5.1 Method

The model takes values on SD scales as its input (hereinafter referred to as the "target"), and the 43 types of scales were used as the SD scales. The model generates SSWs that give an impression designated by the values of the 43 SD scales to the listener.

The 43 scales express various impressions, such as scales expressing an impression of a haptic senses ("hot-cold," etc.), scales expressing an impression of visual senses ("bright-dark," etc.), and scales expressing an impression of subjective senses ("beautiful-dirty," etc.). Table 10.3 lists all 43 scales.

For the purpose of generating novel SSWs with target impressions, we use a genetic algorithm (GA) instead of an existing dictionary-based method [42]. There are two advantages in using a genetic algorithm. First, genetic algorithms are efficient in solving a vast number of combination problems, such as NP-hard optimization. Also, genetic operations such as selection, crossover, and mutation in a GA can be helpful for generating new SSWs from existing ones. Considering that the possible number of combinations of phonemes can be huge, a GA is adequate for the purpose.

As explained, in general a Japanese SSW consists of 2–4 morae, each of which is made up of a combination of a vowel and/or a consonant. Taking this structure into consideration, we designed a phenotype with 17 genes for GA as seen in Table

Table 10.3 List of SD scales

Bright-dark	Fresh-stale	Manly-womanly
Warm-cold	Natural-artificial	Elastic-inelastic
Thick-thin	Familiar-unfamiliar	Glossy-lackluster
Comfortable-uncomfortable	Wet-dry	Strong-weak
Good-bad	Keen-mild	Resistant-unresistant
Impressive-unimpressive	Heavy-light	Bumpy-flat
Happy-sad	Elegant-vulgar	Smooth-rough
Relieved-uneasy	Firm-fragile	Stretchy-rigid
Pleasant-irritating	Simple-complicated	Intense-calm
Hard-soft	Like-dislike	Flashy-modest
Regular-irregular	Slippery-sticky	Cheerful-cheerless
Clean-dirty	Sharp-dull	Western-Japanese
New-old	Static-dynamic	Youthful-aged
Luxury-cheap	Unpolished-buffed	
Eccentric-ordinary	Joyful-doleful	

10.4. Each of them represents a feature of sound or a structural element and takes an integer between 0 and 9.

In the table, the first gene stands for the number of morae of the SSW to be generated. For example, if the value of this gene is set to 3, the SSW to be generated consists of 1 mora. The second gene is a flag of existence of a reduplicated structure. If the value is 1, the SSW will have a mora reduplication. The third gene stands for the consonant of the first mora, which can be one of eight sounds (/k/, /s/, /t/, /n/, /h/, /m/, /y/, /r/, or /w/). The fourth and the fifth genes are flags for voiced sounds and contracted sound. If the fourth flag is true, the consonant will be changed to a voiced sound (e.g., /k/ changes to /g/). If the contracted flag is true, the vowel of the mora will have the remaining pronunciation from the previous character and be recognized as a combination together. The sixth gene stands for the vowel of the first mora, which is one of the five vowels /a/, /i/, /u/, /e/, or /o/. The seventh, eighth, and ninth genes indicate the presence (or absence) of a long vowel, geminate, and syllabic nasal sound, respectively. In the same manner, the rest of the genes except for the fourteenth are used to characterize the second mora. If the fourteenth gene is true, the sound /Li/ is added to the end of the SSW. For instance, given a gene sequence of [7, 1, 2, 2, 6, 5, 0, 3, 2, 8, 6, 3, 4, 8, 6, 9, 1], the SSW is determined to be "Shururin." {(The number of base morae = 2, Reduplication = False, Consonant (1st mora) = s, Voiced sound (1st mora) = False, Contracted sound = True, Vowel (1st mora) = u, Long vowel (1st mora) = False, Choked sound (1st mora) = False, Syllabic nasal (1st mora) = False, Consonant (2nd mora) = r, Voiced sound (2nd mora) = True (But ignored as Japanese /r/ sound does not have a voiced form), Contracted sound = False, Vowel (2nd mora) = u, Long vowel (2nd mora) = True, Choked sound (2nd mora) = True, Syllabic nasal (2nd mora) = False.}

Table 10.4 SSW phenotype architecture for the genetic algorithm

Index	Role	Value and corresponding meaning									
		0	1	2	3	4	5	6	7	8	9
1	Number of morae	1	1	1	1	1	2	2	2	2	2
2	Reduplication	0	0	0	0	0	1	1	1	1	1
3	Consonant (1st mora)	–	k	s	t	n	h	m	y	r	w
4	Voiced sound (1st mora)	0	0	0	0	0	1	1	1	1	1
5	Contracted sound (1st mora)	0	0	0	0	0	1	1	1	1	1
6	Vowel (1st mora)	a	a	i	i	u	u	e	e	o	o
7	Long vowel (1st mora)	0	0	0	0	0	1	1	1	1	1ho
8	Choked sound (1st mora)	0	0	0	0	0	1	1	1	1	1
9	Syllabic nasal (1st mora)	0	0	0	0	0	1	1	1	1	1
10	Consonant (2nd mora)	–	k	s	t	n	h	m	y	r	w
11	Voiced sound (2nd mora)	0	0	0	0	0	1	1	1	1	1
12	Contracted sound (2nd mora)	0	0	0	0	0	1	1	1	1	1
13	Vowel (2nd mora)	a	a	i	i	u	u	e	e	o	o
14	Ending in Li (2nd mora)	0	0	0	0	0	1	1	1	1	1
15	Long vowel (2nd mora)	0	0	0	0	0	1	1	1	1	1
16	Choked sound (2nd mora)	0	0	0	0	0	1	1	1	1	1
17	Syllabic nasal (2nd mora)	0	0	0	0	0	1	1	1	1	1

10.5.2 Generating SSWs by Genetic Algorithm

The brief procedure of generating SSWs with a GA is following:

STEP1: Initialize phenotypes with a specific condition.
STEP2: Generate SSWs from the phenotypes of the current generation.
STEP3: Calculate values on SD scales for the SSWs.
STEP4: Compare the values with the target values. Then selection is to be made by selecting several phenotypes whose values are close to the target as candidates for the next generation.
STEP5: Update selected phenotypes by mutation, and crossover at fixed rate.
STEP6: Repeat *STEP2* to *STEP5* until the given condition becomes true.

Hundreds of phenotypes are prepared and then initialized with the values of SSWs frequently used in daily life at *STEP1*. As these SSWs are randomly selected, decoding of initial phenotypes at *STEP2* results in generating a wide variety of SSWs. Thus, it is necessary to select SSWs with the closest impression to the target among them as parents for the next generation by converting them into comparable values on SD scales at *STEP3*. For this conversion, we previously developed a multiple regression model that outputs the score for each SD scale based on the SSW phonemes, considering the order and other special sounds (such as voiced sounds and contracted sounds) [43]. This regression model was developed based on the results of psychological experiments on sound-symbolic relationships. Specifically, the regression equation is given by Eq. (10.4).

$$SD_i(x) = c1_i(x) + v1_i(x) + \overline{c2_i(x)} + \overline{v2_i(x)} + \overline{m_i(x)} + t_i(x) + r_i(x) + C_i$$
(10.4)

where $SD_i(x)$ denotes the ith SD value for x, $c1_i(x)$ denotes the ith category score for the consonant of the first mora of x, $v1_i(x)$ denotes the ith category score for the vowel of the first mora of x, $\overline{c2_i(x)}$ denotes the average value of the ith category scores for the consonants of the second and later morae of x, $\overline{v2_i(x)}$ denotes the average value of the ith category scores for the vowels of the second and later morae of x, $\overline{m_i(x)}$ denotes the average value of the ith category scores for the special phonological symbols inserted into the middle of x, $t_i(x)$ denotes the ith category score for the special phonological symbol of the last mora of x, $r_i(x)$ denotes the ith category score for whether x is an iteration structure, C_i denotes the value of the ith constant term.

In Step 4, a similarity between a target and generated SSWs is compared by cosine similarity, as given by Eq. (10.5).

$$s(v(x), g) = \frac{v(x) \cdot g}{|v(x)| \, |g|}$$
(10.5)

where g and $v(x)$ denote a vector representation of the target and vector representation of the SSW impression converted by the multiple regression model, respectively. In the default setting of our application, two phenotypes of the largest cosine similarity are taken as the parents for the next generation.

At *STEP 5*, a new generation is to be made by crossover and mutation operations based on selected phenotypes. First, crossover is a genetic operation used to combine the genetic information of two parents for generating new descendants. We adopted single-point crossover in our application, though there exist several popular techniques for crossover operation, such as two-point and k-point crossover. With single-point crossover, a parent's genes are randomly divided into two parts at a crossover point, and the right of the point (the head is at the left) is swapped between the genes, resulting in generating two new descendants. This procedure is repeated until the number of new generated individuals become the same as the number of old generations. Second, to ensure the genetic diversity of the phenotypes of the new generation, another genetic operation called mutation was carried out.

In general, this operation replaces one or more gene(s) of a phenotype from its original value with a designated rate. As it is known that a higher rate of mutation leads to a local optimal solution, the rate of mutation of a gene is typically set between 0.1% and 1%. However, considering that our phenotypes have a very short length of genes, a lower rate of mutation could result in generating identical genes. To maintain the genetic diversity in a generation, we set the mutation rate to 5%. This

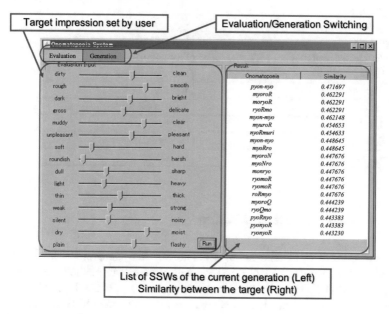

Fig. 10.6 Software appearance

generational process is repeated until one of the termination conditions is achieved: $s(\nu(x), g) > 0.9$, or reaches to 1,000 repetitions).

10.5.3   Software Implementation

We developed the SSW generator as a Windows application with Java. The software is made up of three components. Figure 10.6 shows the appearance of the developed software, where (1) A target impression is given in the left pane, where the user designates values for each of 43 SD scales using slide bars. (2) The switching function between evaluation and generation is given at the tab button at the top. With the evaluation mode, an SSW given to the software is converted into its values on 43 SD scales. All phenotypes in a generation satisfying the termination condition are coded into SSWs and listed by cosine similarity. With the generation mode, selection, crossover, and mutation operations will be repeatedly carried out onto initialized phenotypes according to the procedure described in the previous section. (3) On the pane at the bottom right, a list of SSWs in the current generation is shown with the similarity with the given target vector.

Table 10.5 List of 16 SD scales

Bright-dark	Warm-cold
Comfortable-uncomfortable	Pleasant-irritating
Hard-soft	Luxurious-cheap
Familiar-unfamiliar	Wet-dry
Like-dislike	Slippery-sticky
Joyful-doleful	Manly-womanly
Elastic-inelastic	Glossy-lackluster
Bumpy-flat	Smooth-rough

10.5.4   Performance Evaluation

We performed two experiments to evaluate the SSW generator. The first was an experiment to examine how close SSWs generated by our method are to the target (Experiment A). Specifically, we asked the subjects to input the desired values to the generator and to subjectively evaluate whether the generated SSW was close to the target. The second was an experiment to examine the effect of the initial conditions of the phenotypes for generating SSWs with the desired impression (Experiment B). Two different groups of subjects participated in each experiment.

10.5.4.1   Procedure

In Experiment A, the initial phenotype population was set to 312. Since it was found in prior experiments that the similarity of the best individual can reach 0.9 when repeating the operation about 1,000 times, the number of repetitions for the termination condition of the generation process in the experiment was set to 1,000 times, while for the parameters of the genetic manipulation, the crossover probability was 100% and the mutation probability was 5%. To lighten the workload of subjects, 16 scales out of the original 43 scales considered to be strongly related to the visual and tactile senses were selected (see Table 10.5) and used in the experiment, since these scales are relatively easy for subjects to handle.

First, the subjects were asked to freely input a target impression by changing eight sliders without limitation. Once a target impression was set, the software executed the generation process and then presented the SSW with the largest similarity to the subject, who was asked to evaluate the SSW on a scale of 1 to 7, where a higher score means the generated SSW evoked the most consistent impressions of the target. Subsequently, the subject was also asked to sort the three SSWs with the largest similarity in descending order of subjective similarity. This series of procedures was performed five times for each subject.

In Experiment B, three different initial conditions were compared: (1) initializing all 312 phenotypes with SSWs frequently used in daily conversation, that is, the same initial condition as in Experiment A. (2) Initializing half of the 312 phenotypes with SSWs frequently used in daily conversation. The rest of the phenotypes were initialized with random values. (3) Initializing all 312 phenotypes with random values. The parameters for genetic operation and the termination condition were the same as in Experiment A. Each subject was asked to arbitrarily select three SSWs from their knowledge. For each SSW that presented by a subject, we conducted an experiment with the following procedure.

First, the given SSW was converted into 16 SD scales by the software set to the evaluation mode. The subject was asked to change the SD scales to enhance one (or more) aspect(s) of the impression by manipulating the sliders. In this procedure, the subject was allowed to move the slide bars in a direction "outside" the original value. The generation process by a GA was then executed using these values as an input. Finally, a subject is asked to evaluate a generated SSW from two perspectives: an "impression score" and a "novelty score." (whether the impression of the generated one is close to that intended by the subject), and a "score of novelty" (whether the sound of the generated SSW is novel to the subject) on a scale of 1–7, where a higher score means closer to the target, or sounds novel to a subject.

10.5.4.2   Study Participants

In Experiment A, 18 subjects aged 21–29 years (12 men and 6 women) participated, while 15 subjects aged 20–25 years (8 men and 7 women) participated in Experiment B. All participants were students in the University of Electro-Communications and voluntarily participated in this study.

10.5.4.3   Results

In Experiment A, we obtained 90 (5 trials × 18 subjects) answers. The average score and the standard deviation were 4.54 and 1.74, respectively. According to the analysis of all 90 answers, subjects ranked the SSWs with the highest degree of similarity as the closest to the target impression among the three SSWs presented 35 times in 90 trials, while those of the second and third highest were chosen 30 and 25 times, respectively. These results showed that our strategy for generating SSWs with a desired impression worked relatively well. Also, it was found that the higher the cosine similarity, the more likely it is to match the target impression.

The subjective evaluations of the participants obtained in Experiment B were aggregated and the average value was calculated for each of the three initial conditions. Under the three conditions, the ratios of SSWs in daily use at the time of phenotype initialization were 0%, 50%, and 100%. Figure 10.7 shows the average score of impression/novelty with respect to the initial condition of phenotypes.

As shown in the figure, the average score of impression increases as the proportion of individuals initialized by familiar SSWs increases. Contrary to the tendency of the impression score, the average score of novelty increases as the proportion of

Fig. 10.7 Performance of the GA-based SSW generator under three different initial conditions

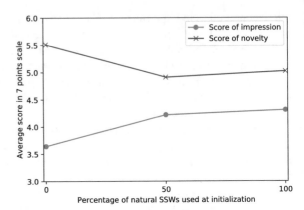

randomly initialized individuals increases. Analysis of variance for each condition for the two types of scores found that the differences of both scores were significant for each condition: $F(2, 176) = 4.45$, $p < 0.05$ (impression), and $F(2, 176) = 3.46$, $p < 0.05$ (novelty).

In summary, we found that using more SSWs commonly used in daily life at initialization produces SSWs more familiar to the subjects. On the other hand, the experimental results also showed that by using completely random values at initialization, combinations of phonemes and morphologies that are not known to the subject are likely to be obtained, resulting in more novel SSWs being generated. The results of the evaluation experiments also suggest that the proposed method is effective as a technology that supports the creation of novel SSWs. The present method can only generate fixed-length SSWs. Therefore, in future work, we will expand the present method to arbitrarily control the length of the SSWs to be generated.

10.6 Machine Tells Us New Potential Values

Let us introduce an example in Japan. There are many things with soft textures such as blankets, bread, rice cakes, facial tissue, cotton, slime, and baby skin. There are many SSWs expressing soft textures, such as "fuwa-fuwa," "fuka-fuka," "howa-howa," "honya-honya," "hunya-hunya," "mowa-mowa," "moko-moko," and "mochi-mochi." There would seem to be more than enough SSWs, but at one point someone started using a new SSW "mofu-mofu" to express the texture of the hair of cats and dogs, and people began using this word. There are various theories about the time when "mofu-mofu" was created in the first place, and it is difficult to identify who first used the word when and where, but it seems that the word was first used in comics. As far as we know, for example, in the 14th volume (2001) of *Shaman King* by Hiroyuki Takei, "mofu-mofu" is used as an expression for the sound of eating bread or some characteristic of bread. The use of "mofu-mofu" for animals, which

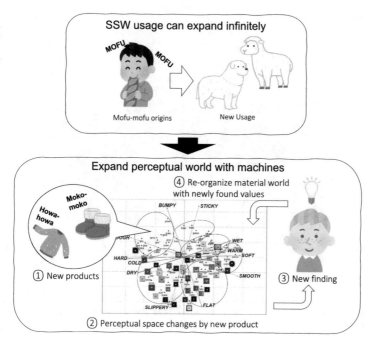

Fig. 10.8   Machines tell us new potential values

is now mainstream, was around 2003 or 2004 on the Internet, for example, "mofu-mofu: Let's play with cute Mofu," and finally it became popular among female junior and senior high school students in 2010. A search for "mofu-mofu" now shows that it is more often used for animals than for bread. The word would have been a perfect fit to convey some characteristics of animals that cannot be conveyed by existing linguistic expressions, and now the new texture word has become established among the Japanese people (Fig. 10.8).

This new texture word plays an important role in our society, letting us find a new attractive value of certain cat hair textures. If the texture of the cat's hair were expressed by one of the existing SSW, for example, the SSW "fuwa-fuwa" that is often used to represent brushed objects such as ordinary blankets, cat's hair and blankets would be in the same category. People could not recognize the characteristics and unique attractiveness of cat hair. With the birth of the new SSW "mofu-mofu," not only softness but also warmth, slower movements, and cuteness were emphasized and a new fashion was born, such as the popularity of animal video special features called "mofu-mofu videos." Sometimes people spontaneously create new words and create new values, but machines can facilitate this process. We believe that the AI that generates the new SSW introduced in Sect. 10.5 could yield a new value.

If new SSWs are located in the word-based perceptual space introduced in Sect. 10.4, the word-based perceptual space might be used for product design (for an example of fabric texture design inspired by SSWs, see [44]) and communica-

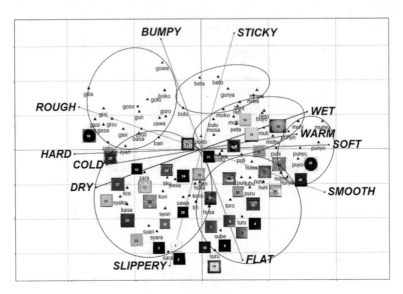

Fig. 10.9 Distribution diagram of 40 tactile materials. The diagram indicates that the category including "gori-gori," "goso-goso," and "gisi-gisi" and the category including "gunya-gunya," "beto-beto," and "beta-beta" lack corresponding materials

tion between designers and customers in the field of product design. If a customer requests a texture using a new SSW located on the diagram, a designer can develop the new material/product by referring to the location of the SSW, and/or to an existing material/product located nearby on the diagram. If this kind of visualization is used for a specific product, such as cosmetics or fabrics, only a collection of words is needed to make a new diagram suitable for the product. After collecting SSWs used to express the product, the words are automatically converted into rating scores of 26 adjectives by the system, and a new word-based diagram can be constructed.

To develop 50 materials that have a wide range of texture qualities, materials that currently exist can be placed on the diagram to reveal what kind of texture is lacking. If a collection of 40 materials is placed as in Fig. 10.9, it is obvious that the category including "gori-gori," "goso-goso," and "gisi-gisi" and the category including "gunya-gunya," "beto-beto," and "beta-beta" are needed for the other 10 materials, as in Fig. 10.10.

When a new texture expression such as "mofu-mofu" is born, there is a possibility that a new material will also be born based on that newly created expression, and when a new object is born, a new word will be born to name the object. In general, it is a human being that happens to notice a new texture that cannot be expressed by existing words and creates words that express that texture. If a machine reveals a new texture for which a texture expression can be generated using the system introduced in Sect. 10.5, this might make people aware of the potential value of a material described by that expression, new products will be born, and the physical world will change as well.

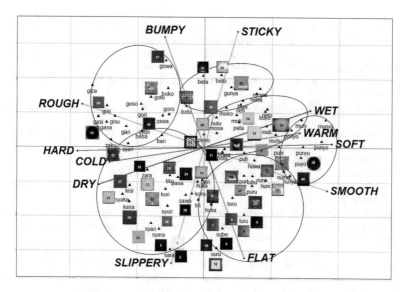

Fig. 10.10 Distribution diagram of 50 tactile materials, with 10 materials added to Fig. 10.9

Acknowledgements This work was supported by a Grant-in-Aid for Scientific Research on Innovative Areas "Shitsukan" (No. 23135510 and 25135713) from MEXT Japan and JSPS KAKENHI Grant Number 15H05922 (Grant-in-Aid for Scientific Research on Innovative Areas "Innovative SHITSUKSAN Science and Technology") from MEXT, Japan.

References

1. Le, Q.V. et al.: Building high-level features using large scale unsupervised learning (2011). arXiv:1112.6209
2. Jones, L.A., Lederman, S.J.: Human Hand Function. Oxford University Press (2006)
3. Lederman, S.J., Klatzky, R.L.: Haptic perception: a tutorial. Atten. Percept. Psychophys. **71**, 1439–1459 (2009)
4. Bensmaia, S.J.: Texture from touch. Scholarpedia **4**, 7956 (2009)
5. Tiest, W.M.B.: Tactual perception of material properties. Vision. Res. **50**, 2775–2782 (2010)
6. Whitaker, T.A., Simões-Franklin, C., Newell, F.N.: Vision and touch: independent or integrated systems for the perception of texture? Brain Res. **1242**, 59–72 (2008)
7. Tamura, H., Mori, S., Yamawaki, T.: Textural features corresponding to visual perception. IEEE Trans. Syst. Man Cybern. **75**, 460–473 (1978)
8. Lederman, S.J., Thorne, G., Jones, B.: Perception of texture by vision and touch: multidimensionality and intersensory integration. J. Exp. Psychol. Hum. Percept. Perform. **12**, 169–180 (1986)
9. Sucevic, J., Jankovic, D., Kovic, V.: When the sound-symbolism effect disappears: the differential role of order and timing in presenting visual and auditory stimuli. Psychology **4**(7A), 11–18 (2013)
10. Dingemanse, M., Blasi, D.E., Lupyan, G., Christiansen, M.H., Monaghan, P.: Arbitrariness, iconicity, and systematicity in language. Trends Cognit. Sci. **19**(10), 603–615 (2015). https://doi.org/10.1016/j.tics.2015.07.013

11. Asano, M., Imai, M., Kita, S., Kitajo, K., Okada, H., Thierry, G.: Sound symbolism scaffolds language development in preverbal infants. Cortex **63**, 196–205 (2015)
12. Maurer, D., Pathman, T., Mondloch, C.J.: The shape of boubas: sound-shape correspondences in toddlers and adults. Dev. Sci. **9**, 316–322 (2006)
13. Westbury, C.: Implicit sound symbolism in lexical access: evidence from an interference task. Brain Lang. **93**, 10–19 (2004)
14. Ohala, J.J.: Sound symbolism. In: Proceedings of 4th Seoul International Conference on Linguistics, pp. 98–103 (1997)
15. Ramachandran, V.S., Hubbard, E.M.: Synaesthesia-a window into perception, thought and language. J. Conscious. Stud. **8**(12), 3–34 (2001)
16. Jespersen, O.: Symbolic Value of the Vowel I. Publisher unknown (1921)
17. Köhler, W.: Gestalt psychology. H. Liveright, New York (1929)
18. Sapir, E.: A study in phonetic symbolism. J. Exp. Psychol. **12**(3), 225–239 (1929). https://doi.org/10.1037/h0070931
19. Bremner, A., Caparos, S., Davidoff, J., de Fockert, J., Linnell, K., Spence, C.: Bouba and Kiki in Namibia? A remote culture make similar shape-sound matches, but different shape-taste matches to Westerners. Cognition **126**, 165–172 (2013)
20. Brown, R.W., Black, A.H., Horowitz, A.E.: Phonetic symbolism in natural languages. J. Abnorm. Soc. Psychol. **50**(3), 388–393 (1955). https://doi.org/10.1037/h0046820
21. Davis, R.: The fitness of names to drawings: a cross-cultural study in Tanganyika. Br. J. Psychol. **52**(3), 259–268 (1961). https://doi.org/10.1111/j.2044-8295.1961.tb00788.x
22. Emeneau, M.B.: Onomatopoetics in the Indian linguistic area. Language **45**(2), 274–299 (1969). https://doi.org/10.2307/411660
23. Enfield, N.J.: Areal linguistics and mainland southeast Asia. Ann. Rev. Anthropol. **34**, 181–206 (2005)
24. Hinton, L., Nichols, J., Ohala, J.: Sound Symbolism. Cambridge University Press, Cambridge, UK (1994)
25. Klank, L.J.K., Huang, Y.-H., Johnson, R.C.: Determinants of success in matching word pairs in tests of phonetic symbolism. J. Verbal Learning Verbal Behav. **10**(2), 140–148 (1971). https://doi.org/10.1016/S0022-5371(71)80005-1
26. Kovic, V., Plunkett, K., Westermann, G.: The shape of words in the brain. Cognition **114**, 19–28 (2010)
27. Nuckolls, J.B.: The case for sound symbolism. Ann. Rev. Anthropol. **28**, 225–252 (1999)
28. Schmidtke, D.S., Conrad, M., Jacobs, A.M.: Phonological iconicity. Front. Psychol. **5**(80), 1–6 (2014)
29. Voeltz, F.K.E., Kilian-Hatz, C. (eds.): Ideophones. John Benjamins, Amsterdam (2001)
30. Bloomfield, L.: Language. Henry Holt (1933)
31. Bolinger, D.: Rime, assonance and morpheme analysis. Word **6**, 117–136 (1950)
32. Crystal, D.: The Cambridge Encyclopedia of the English Language. Cambridge University Press (1995)
33. D'Onofrio, A.: Phonetic detail and dimensionality in sound-shape correspondences: Refining the bouba-kiki paradigm. Lang. Speech **57**(3), 367–393 (2013). https://doi.org/10.1177/0023830913507694
34. Gallace, A., Boschin, E., Spence, C.: On the taste of "bouba" and "kiki": an exploration of word-food associations in neurologically normal participants. Cognit. Neurosci. **2**, 34–46 (2011)
35. Sakamoto, M., Watanabe, J.: Cross-modal associations between sounds and drink tastes/textures: a study with spontaneous production of sound-symbolic words. Chem. Senses **41**(3), 197–203 (2016)
36. Sakamoto, M., Watanabe, J.: Bouba/kiki in touch: associations between tactile perceptual qualities and Japanese phonemes. Front. Psychol. **9**(295), 1–12 (2018). https://doi.org/10.3389/fpsyg.2018.00295
37. Hamano, S.: The Sound-Symbolic System of Japanese. CSLI Publications, Stanford, CA, Tokyo, Kuroshio (1998)

38. Okamoto, S., Nagano, H., Yamada, Y.: Psychophysical dimensions of tactile perception of textures. IEEE Trans. Haptics **6**(1), 81–93 (2013)
39. Sakamoto, M., Watanabe, J.: Effectiveness of onomatopoeia representing quality of tactile texture: a comparative study with adjectives. In: Papers from the 13th National Conference of the Japanese Cognitive Linguistics Association, pp. 473–485 (2013). (in Japanese)
40. Doizaki, R., Watanabe, J., Sakamoto, M.: Automatic estimation of multidimensional ratings from a single sound-symbolic word and word-based visualization of tactile perceptual space. IEEE Trans. Haptics **10**(2), 173–182 (2016). https://doi.org/10.1109/TOH.2016.2615923
41. Sakamoto, M., Watanabe, J.: Visualizing individual perceptual differences using intuitive word-based input. Front. Psychol. **10**(1108), 1–8 (2019). https://doi.org/10.3389/fpsyg.2019.01108
42. Holland, J.H.: Genetic algorithms and the optimal allocation of trials. SIAM J. Comput. **2**(2), 88–105 (1973). https://doi.org/10.1137/0202009
43. Sakamoto, M.: System to quantify the impression of sounds expressed by onomatopoeias. Acoust. Sci. Technol. **41**(1), 229–232 (2020). https://doi.org/10.1250/ast.41.229
44. Hemmings, J.: NUNO Books (NUNO Corporation), Posted on Wed, January 1st, 2003 in Book Reviews (2003). http://www.jessicahemmings.com/nuno-books-nuno-corporation/. Accessed from 27 June 2016
45. Sakamoto, M., Watanabe, J.: Visualizing individual perceptual differences using intuitive word-based input. Front. Psychol. **10**, 1108 (2019)

Chapter 11
Interactive Sensing and Sensing Interactions

Eri Sato-Shimokawara

Abstract Sensors are collecting data from various environments, various situations, not only machinery but also human beings, and these data have been employed for the basis of various services for general/individual purposes. Progress of IoT technology encourages collect and use of data under ordinary life, it's mean the system obtains a chance to adapt human/community/localization. This chapter focused on sensing humans activity and describes the difficulties of acquiring eligible data, and shows the effectiveness of interaction.

11.1 Sensing Living Things and Things to Live With

Progress in the IoT has created a variety of sensing technologies that can be used for monitoring, tracking, and logging human activities. These facilitate digital collections called lifelogs. Jacquemard et al. [10] reviewed Lifelog technologies from the perspectives of challenges and opportunities, and divided them into four areas of use: private, corporations, public institutions, and governmental organizations. They also discussed the ethical debate over the potential domain of application. Therefore, the ethical part is not discussed in this chapter.

This section focuses on lifelog care in private settings such as the home from the viewpoint of health care. Because health support and care are of serious interest not only for the elderly but for everyone in the current era of the 100-year life [5], a modern approach is to support healthy (both physically and mentally) seniors in living alone, but it is hard to keep healthy by oneself. This section introduces the lifelog collection method for the monitoring of a user's life to providing suitable services to maintain health. As such challenges have been widely researched, this chapter focuses only on daily activities. Traditional monitoring systems have focused on elementary motions such as walking and sleeping. These motion data are easily collected by wearable devices such as a smartphone or smart-watch, which are useful to support self-health care and maintain regular life rhythms. However, in daily life,

E. Sato-Shimokawara (✉)
Tokyo Metropolitan University, 6-6 Asahigaoka, Hino, Toyo, Japan
e-mail: eri@tmu.ac.jp

© Springer Nature Switzerland AG 2023
Y. Ohsawa (ed.), *Living Beyond Data*, Intelligent Systems Reference Library 230,
https://doi.org/10.1007/978-3-031-11593-6_11

our motions are very varied. Therefore, this section introduces research on methods for estimating the basic Activities of Daily Living (ADLs) normally performed in daily life such as feeding, bathing, and homemaking. This study selects 11 motions associated with ADLs for recognition. We deal with the problem of recognizing advanced motions by using transportation data for movements from one area to other areas in a house. These advanced motion logs support not only the elderly but also the care staff or family who plans the support.

Currently a variety of motion capture techniques are being researched and developed. The most popular detection tools use image devices such as Kinect or Xtion. However, the area that these devices capture is too narrow to observe users' motions in daily life. Therefore, this research deals with recognizing motion by using 3D acceleration sensors with ZigBee communication. 3D acceleration sensors are very popular sensors embedded in smart phones. Activity recognition using the accelerometer has been investigated by many researchers. Bao and Intille [2] researched the recognition of 20 activities by 5 accelerometer sensors, showing that accelerometer data can be used to classify activities with an accuracy of around 80%. Chen et al. [3] developed activity monitors using an accelerometer to predict energy expenditure. Ravi et al. [16] studied the recognition of eight activities by an accelerometer worn near the pelvic region. Other research is surveyed in [15]. These researchers reported that the accuracy of the recognition is roughly proportional to the number of sensors. In this research, we considered how to use as few sensors as possible for practical use.

Ouchi and Doi [13] developed an activity recognition system by using the accelerometer sensor and microphone on mobile phones. This system is an appropriate device for studying activities that make sounds. However, not all our activities make sounds and we do not always have a mobile phone on hand. In a previous study, we succeeded in recognizing seven motions: lying down, sitting, standing, walking, running, getting up, and falling down, using only one 3D acceleration sensor [9]. This paper focuses on location because the activities of daily living include motions that are difficult to distinguish using only accelerometer data. Some of our daily life motions are linked to locations; for example, we wash our face at the lavatory and dishes in the kitchen. Hence, this research uses the received signal strength indication (RSSI) to detect the location of the user.

11.1.1 System Outline

Figure 11.1 shows an overview of our system. This study uses a 3D acceleration sensor and ZigBee network. Acceleration data are sent to a server via an access point or coordinator. The server estimates the user's ADL motion using area recognition by RSSI and motion recognition with the acceleration data. The next subsection shows the details of the hardware and software composition.

ZigBee accelerometers made by System Craft Inc. is used to recognize the user's conditions. It consists of a coordinator, access points, and sensor devices. Access

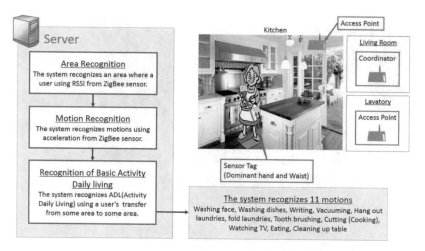

Fig. 11.1 Overview: Proposed system recognizes 11 Activities of Daily Living (ADLs) using area recognition and motion recognition. Each recognition uses the ZigBee sensor network

points are placed in each room and the sensor tag devices are attached to the user's body. The next section discusses how many sensors are needed and the best places to attach the sensor to the user's body. These devices construct a sensor network that records 3D acceleration data 20 times per second and position data 24 times per minute from sensor devices inside the valid range of the ZigBee network. This device can be used to obtain the signal strength indication (RSSI), which is useful for detecting the device position, because RSSI reflects the distance between the sensor tag device and an access point (or coordinator) [4]. Each access point has approximately 20 m of radio range. In a previous study, the authors developed motion recognition algorithms for this sensor network [9] that allow this sensor to recognize the user's position, posture, walking conditions, and whether they have fallen down or are getting up. A system diagram of the 3D accelerometer sensor network is shown in Fig. 11.2.

Figure 11.3 shows a flowchart of the process of recognizing ADL motions. First, the system detects the location of the user using RSSI values. Next, the system selects a fuzzy rule using the result of area recognition. Finally, the system recognizes the motion using acceleration data with the selected fuzzy rule. The recognition method based on fuzzy inference is described in Sect. 11.1.3.

11.1.2 Area Recognition

We conducted an experiment to determine whether RSSI values can be used for area recognition and the best location on the user's body to attach the sensors. The subjects were five students in their 20s and one man in his 40s.

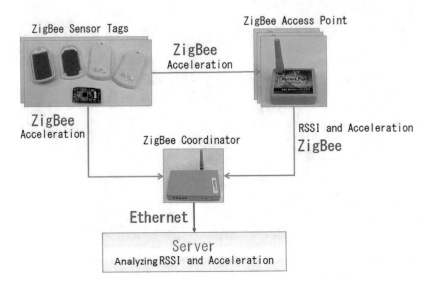

Fig. 11.2 The ZigBee Sensor network consists of sensor tags, access points, and a coordinator. The server obtains the RSSI and acceleration data from the coordinator

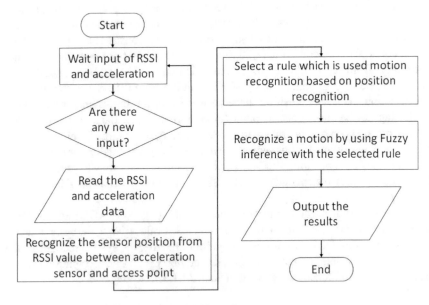

Fig. 11.3 Flowchart of the recognition of ADL motions

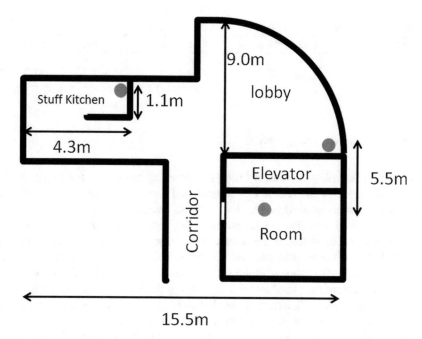

Fig. 11.4 Pre-experiment map: • is an access point or a coordinator

Table 11.1 Result of area recognition

Area	Recognized area (times)			Recognition rate (%)
	Lobby	Room	Kitchen	
Lobby	213	0	40	84.1
Room	0	270	0	100.0
Kitchen	1	0	266	99.6

This section describes the accuracy of area recognition using RSSI values, which change with the distance between the sensor and access point (or coordinator). However, as RSSI is affected by obstacles like walls, RSSI does not always change linearly with the distance. Figure 11.4 shows a map of the experiment location. We established two access points on the ceiling in the staff kitchen and lobby, and the coordinator is established in the room. • shows the access points and coordinator.

Subjects attached a sensor to their waist and stood under (or near) the access point. We collected data for 10 min. Table 11.1 shows the result of this experiment. The recognition rate in the lobby was low because there was a wall between the lobby and staff kitchen. However, the recognition rates were all over 84%, and the average was 94.8%.

Table 11.2 Comparison of sensor tag numbers and usable data

Tag number	Collected data	Usable data	Rate (%)
2 tags	8971	8562	95.4
5 tags	18001	11589	64.4

11.1.3 Activity Recognition

This section describes the best position for attaching the sensor tag to the user's body. We compared the use of two sensors and five sensors. In the case of two sensors, the sensors were attached to the dominant hand and waist. In the case of five sensors, they were attached to the right hand, left hand, right ankle, left ankle, and waist. The experiment results are presented in Table 11.2, showing that data were frequently lost when the user attached five sensors. One reason is that the receiving speed was too low to capture all the sensor data. We thus used only two sensors, attached to the dominant hand and waist (center of the body) as shown in Fig. 11.5. For the two sensors, the x-axis indicates left-right position, the y-axis is directed in the forward-backward direction, and the z-axis is in the vertical direction.

We tried traditional methods of frequency analysis for the obtained data. Figure 11.6 shows an example of the result using the frequency spectrum of the

Fig. 11.5 Tag position: We set two tags on the subject's body: On the dominant hand and waist

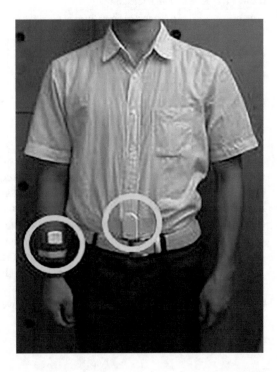

Fig. 11.6 Results of FFT

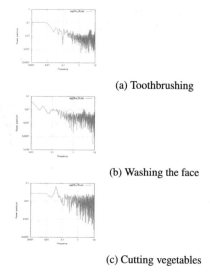

(a) Toothbrushing

(b) Washing the face

(c) Cutting vegetables

x-axis acceleration data for the hand. The result shows that the frequency data include high-frequency content as noise, and traditional frequency analysis is thus unsuitable for this system. Therefore, we approach the system using fuzzy inference. The next section describes the recognition method using fuzzy inference.

Frequency analysis is insufficiently sensitive to estimate the ADLs. Yihshin et al. developed a motion recognition system using fuzzy inference [7]. Therefore, we take an approach using fuzzy inference.

First, the membership function and fuzzy rule are designed based on the average acceleration data, which are recorded every 5 s. The features of activities in the lavatory and kitchen are easily identified. However, the motions in the living room are difficult to determine from the average acceleration, so we tried to use other data as described in the next section. Figure 11.7a, b show examples of membership functions. Equations (11.1) and (11.2) are examples of fuzzy rules; here we let the membership functions of Fig. 11.7a, b be A and B, respectively, where the left side is "Low" and the right side is "High."

$$\text{IF A } is \text{ Low } and \text{ B } is \text{ High } then \text{ Facewashing} \qquad (11.1)$$

$$\text{IF A } is \text{ High } and \text{ B } is \text{ Low } then \text{ Toothbrushing} \qquad (11.2)$$

The motions that occur in the living room are estimated using the average of the resultant vector or ratio of each axis value, because in the living room, the motions do not have movements in a specific direction. Figure 11.8a, b show examples of the membership function based on the average of the resultant vector. The membership function is used for estimating the hand's writing motion. The fuzzy rule is shown in Eq. (11.3). Here, let the right function of Fig. 11.8a, the left function of Fig. 11.8a,

Fig. 11.7 Membership
function for the lavatory

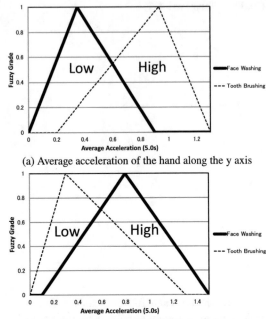

(a) Average acceleration of the hand along the y axis

(b) Average acceleration of the hand along the z axis

the right function of Fig. 11.8b, and the left function of Fig. 11.8b be C, D, E, F, respectively. Figure 11.8c, d show examples of the membership function by the ratio of each axis. The membership function is used for estimating the vacuuming motion. The fuzzy rule is shown in Eq. (11.4). Here, let Fig. 11.8c, d be G and H, respectively.

$$\text{IF } max(C, D) \text{ and } max(E, F) \text{ then Writing} \tag{11.3}$$

$$\text{IF } max(G, H) \text{ then Vacuuming} \tag{11.4}$$

11.1.4 Experiment

We tested the system using the fuzzy inference as described in the above section. The subjects were six students in their 20s. As stated above, the sensors were attached to each subject's dominant hand and waist. The subjects then performed 11 activities. Figure 11.9 shows a correspondence table for motions and locations.

Tables 11.3, 11.4, and 11.5 show the results for the lavatory, kitchen, and living room, respectively. In the living room, we first tried to estimate seven activities.

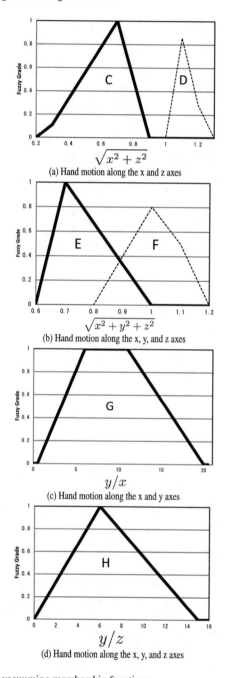

(a) Hand motion along the x and z axes

(b) Hand motion along the x, y, and z axes

(c) Hand motion along the x and y axes

(d) Hand motion along the x, y, and z axes

Fig. 11.8 Writing and vacuuming membership functions

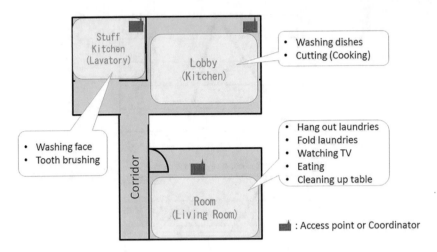

Fig. 11.9 Experiment map

However, it is very difficult to estimate these activities because these activities include motions that are not directed in a specific direction or do not involve specific tools. Therefore, the system classified them into four activities: TV watching, handwriting, vacuuming, and other activities, at the present stage. Overall, the estimation ratio is 80.6–92.7% (Tables 11.3, 11.4, and 11.5).

11.1.5 Conclusion

This section introduced a system that estimates the basic ADLs using 3D acceleration data. We selected 11 activities from among the ADLs for estimation. We established a ZigBee sensor to obtain the acceleration data. The ZigBee sensor helps to narrow down the location of the user. The system estimates the activities using the acceleration data and area recognition. In this research, we used fuzzy inference to recognize the motions. The system selects the fuzzy membership function and the rules according to the results for area recognition. We tested the system with data collected from six participants. Accordingly, we found that estimation is very hard in the living room, so we focused on three activities. The result shows that the estimation ratio is over 80% for about seven activities (Tables 11.3, 11.4, and 11.5).

In the future, this system can be improved to increase the accuracy of estimation by using the movement history of daily life; for example, users often cook at breakfast time. This collection of activity history can serve as the basis for constructing a user model, which is useful for improving the activity estimation accuracy and for enhancing services.

Table 11.3 Estimation results for the lavatory

Motion	Area recognition rate (%)	Motion recognition rate (%)	Motion estimation rate (%)
Washing the face	88.8	99.7	88.4
Toothbrushing	89.9	97.6	89.9

Table 11.4 Estimation results for the kitchen

Motion	Area recognition rate (%)	Motion recognition rate (%)	Motion estimation rate (%)
Cutting(cooking)	90.5	89.7	81.2
Washing dishes	92.8	86.9	80.6

Table 11.5 Estimation results for the living room

Motion	Area recognition rate (%)	Motion recognition rate (%)	Motion estimation rate (%)
Watching TV	100.0	92.7	92.7
Writing	99.1	85.1	84.3
Vacuuming	97.2	83.7	81.4

11.2   Collecting Human Interest from Interaction [17]

Japan is one of the longest-lived populations [12]. Preserving elderly people's physical and mental health to enhance their QOL (Quality of life) is a high-priority issue. However, depression is one of the leading causes of DALYs and YDLs [8]. Conversation is an important factor in preventing or relieving depression. However, the proliferation of nuclear households has led to reduced conversation. Therefore, we focused on chat robots, which have been developed and researched by many researchers. For example, Siri is the well-known natural speech interface released by Apple [11]. Its main task is providing support to control a phone, but users sometimes chat with Siri. The greatest problem in developing a chat robot is creating a sentence database, which some robots collect from Web data such as Twitter or some SNSs. However, these data are not necessarily suitable for enjoyable conversations. Therefore, we have proposed a cloud-based chat robot system [17]. First, the chat robot system selected a sentence from the database using just word matching. Participants noted that the robot changed the topics frequently, which sometimes made it difficult to follow and understand the robot dialogue. We improve the sentence selection method using category estimation. This paper analyzed the validity of category estimation compared with manual methods.

First we explain our robot system and category estimation method. Second, we discuss the validity of topic estimation. Finally, we summarize this study and discuss future research.

11.2.1 Proposed Category Estimation Method for Chat Robot

Topic estimation is researched in the area of natural language processing. Özyurt and Köse proposed a topic determination method for text chats [14]. Their proposed method classified data into six categories using Naive Bayes, k-NN, and SVM, yielding accuracy ratios from 84.5 to 91.7%. However, this method focused on the unique patterns in text chat such as acronyms, short forms, and icons. As our developed chat robot uses speech recognition, we must develop category estimation methods for voice conversations, which are necessary for quick responses to maintain a natural pace. We then proposed the simple method shown in Fig. 11.3.

The proposed method used the results of morphological analysis. Each separated word is checked with the keyword list, which was created by manually using data obtained from an experiment in June 2014. First we selected subjects who used the chat robot frequently from the previous experiment subjects. An annotator classified the 1542 sentences in the subject's dialogue history into 18 categories (see Table 11.6). These 18 categories were obtained when we created a fixed phrase database [20]. We obtained some nouns from the annotated sentences by morphological analysis (Table 11.6 shows part of the keyword list). The proposed estimation method counts words matching the keywords list. If a sentence includes words that match the keyword list, the category with the highest appearance ratio is the category for the sentence. If a sentence does not have any words included in the keyword list, the method uses the robot's sentence. The robot sentence is selected from the dialogue database or fixed phrases, or is created using a template with Wikipedia. The sentences in the dialogue database and fixed phrases form a category established in advance. Thus, if the robot's sentence is selected from the dialogue database or fixed phrases, the category stored in the database serves as the sentence category. In case no pattern is matched, the sentence category is set as "other".

The previous chat system selects a sentence from the database using nouns [17]. This system selects a sentence from a category when the next robot sentence will be selected from a fixed phrase database and the subject sentence is categorized as NOT "other" (Fig. 11.10).

11.2.2 Experiment

We carried out an experiment using this category estimation method with 10 subjects, all over 65 years old. Five subjects had used the chat robot in the previous experiment, which was carried out in June and July 2014. We set up the chat robot in the subjects'

Table 11.6 Part of the keyword list for category estimation

Category	Words (number of including words)
Health	Health, life, illness, tobacco (10)
Environment	Weather, rain, season (5)
Society	Community, war (3)
Music	Music, karaoke, piano (7)
Event	Festival, concert, birthday (4)
Family	Husband, grandchild, brother (11)
Washing	Detergent, fabric softener (6)
Fashion	Shoes, clothes, yukata(Japanese summer kimono) (6)
Go-out	License, car, shopping, outing (21)
Meal	Egg, bread, appetite, fruit (24)
Game	Game center (1)
Work	Job (3)
Media	Mail, internet, phone, TV, book (13)
Machine	PC, camera, phone (7)
Sleeping	Pillow, dream, night (8)
Cleaning	Clean (3)
Exercise	Sports, baseball, outing (9)
Cooking	Tempura, meal (6)

*Number in bracket shows the number of words included in the category

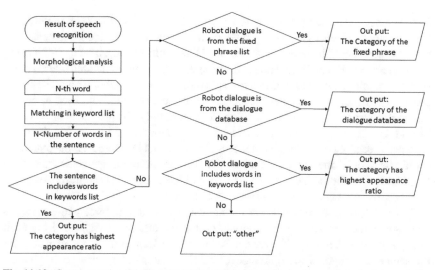

Fig. 11.10 Category estimation flowchart: First, this method checks the words in the keywords list using the results of morphological analysis. Second, this method uses the robot utterance sentence; if the sentence was selected from fix phrase database or dialogue database, this method outputs the category of the database. In a case other than the above, the category is determined as "Other"

home as in the previous study. The experimental period was from December 2014 to February 2015. Each subject used the robot around three weeks. In this experiment, one subject withdrew from the experiment because of mechanical problems. Finally, we obtained a log of 13,727 dialogue sentences, which served as the basis of the analysis in this paper.

11.2.3 Analysis of Category Estimation

This experiment can be analyzed from various perspectives. The effect of category estimation on conversation is one of the important topics, but this consideration is currently under further analysis and will be reported in other papers.

This paper considers the validity of the category estimation method. Therefore we categorized every sentence manually. However, the category selected differed by annotators. Each sentence was then annotated by 3 persons; 10 students participated as annotators.

Section 11.2.3.1 discusses the differences across the three annotators, while Sect. 11.2.3.2 discusses the validation based on a comparison of the manual annotation with the category estimation method.

11.2.3.1 Analysis of Individual Annotator Differences

Each sentence was categorized by three annotators, because for some sentences it was difficult to decide a category. A set of dialogues consisted of a robot sentence and a subject sentence. The target sentences used the results of speech recognition using Android Speech Recognizer. The results of speech recognition include many errors, so the annotator often had to guess the actual data from the robot's sentence or the previous sentence. The annotator checked the subject sentence and tried to decide on the most appropriate of the 18 categories using only a subject sentence ((1) in Fig. 11.11). If it was difficult to decide, the annotator would then check the robot sentence preceding the subject sentence ((2) in Fig. 11.11). Moreover, the annotator could check only one previous subject sentence before the target sentence ((3) in Fig. 11.11). If the annotator decides that none of the categories is appropriate for the sentence, the sentence category is set as "other".

Table 11.7 shows the results for the agreement among the three annotators. No consistency means that all three annotators selected different categories from each other, indicating that it is difficult even for humans to decide on the category because the sentences straddled multiple categories. For example, "Sounds good, you are working as volunteer" was categorized as "Work," "Event," and "Society." The pair of sentences in which the robot said, "I enjoyed lunch time with my friend" and the subject said, "It is good. I sometimes eat with my friend, too," was categorized as "Meal," "Go-out," and "Society."

Utterer	Contents	Category
Robot	宝くじは買ったことありますけど、当ったことがありません。 I had bought lotto, but I had never win.	Event
③Subject	宝くじはそうですよね―ないですね [だから最近会わせる] Lotto…I had never bought it, [then I meet].	
② Robot	昔は海外で暮らしたいと思っていたんですけど、最近はあまり思わないですね。 I thought I want to live foreign country, but now I don't think so.	
① Subject	そうですね テロと[金] 色んな事件が起きて 日本安全かな と思いますね 私も[恐れ]に行きたいなあ と思ったことありますよ 英語 できたらいいなって Well, many accidents are occurred such as terror, [money, and so on]. I think Japan is safety. I thought I want to go to [fear], I hope I can speak English.	Event
Robot	何をするにもお金がかかりますね。 Need money to do something	Other
Subject	金かかりますねぇ 習い事するのに[胸下の痛みるん]ですけどね Yes, money is needed. Taking a culture lesson, [I have ache under my breast].	
Robot	今日は習い事の日だったので、色んな方と色んな話をしました。楽しかったですよ。 I have a culture lesson today, then I talked about many things with many people. It was so fun.	Eating
Subject	そのですね 私も良行お話しますよねw[でさらにきましたね 甲子園飲みながらね] It sounds good. I often talk and [come more with drinking a Koushien].	

Fig. 11.11 Annotation method for subject utterance. [ ] in the contents means error of speech recognition

Therefore, we used 12,783 sentences that were annotated with the same category by 2 or 3 annotators as categorized sentences. Figure 11.12 shows the numbers of categorized and non-categorized sentences for each subject. Table 11.8 shows the ratio of total sentence to categorized sentence. The total number of sentences reflects variability among the subjects, but the ratios is same tendency. However, around 90% of the sentences are categorized (meaning that 2 or 3 persons annotated it with the same category) in each subject's data (Table 11.8).

Next, we focused on the category ratios. The sentences categorized as "Other" account for more than 55% of the total categorized sentences (Fig. 11.3) and include greetings ("'hello," "good morning," etc.), agreement ("I think so too."), and those that are difficult to understand because of the failure of speech recognition. Figure 11.14 shows a breakdown of the sentences classified into the 18 categories. The dialogue log includes many sentences talking about "Go-out" and "Meal"; this result is nearly the same as in the previous experiment [19]. We consider "Go-out" to include the usual behaviors of daily life such as walking around and shopping. Moreover, topics of travel or trips include "Go-out," so these topics are not difficult to talk about in chatting. Meals are also usual behavior; as everybody takes a meal 2 or 3 times a day, subjects easily answered questions such as "What did you eat today for breakfast?" and "What is your favorite food?"

11.2.3.2 Comparison of Manual Annotations with Category Estimations

This section analyses the validation of category estimation by the robot. Developed robot decided to select a category from the 18 categories or "Other" as described in

Table 11.7 Numbers of sentences categorized consistently by manual classification

	3 people	2 people	No consistency
Subject 1	620	564	93
Subject 2	350	331	44
Subject 3	853	1035	111
Subject 4	83	39	3
Subject 5	99	151	28
Subject 6	701	787	93
Subject 7	794	1542	320
Subject 8	291	332	75
Subject 9	1938	1308	109
Subject 10	384	581	68
Total	6113	6670	944

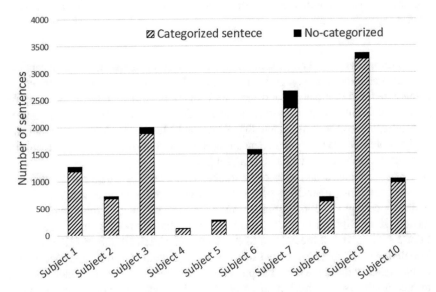

Fig. 11.12 The number of manually categorized and non-categorized sentences for each subject

Fig. 11.10. The decided category is invoked as a control to select a sentence from the dialogue database and is stored log database with the sentence. The cloud server manager chooses sentences that can be reused as robot dialogue and updates the dialogue database with the category recognized by the robot automatically.

Therefore, this section compares the categories as classified by the robot and humans. Figure 11.15 shows the results. The basic data comprise 5671 acceptable sentences categorized as "Other" from among the total categorized sentences. Coincidence means that the robot categorized the sentence the same as the human annotator.

Table 11.8 Numbers and ratios of categorized sentences

	Total	Categorized	Ratio [%]
Subject 1	1277	1184	92.7
Subject 2	725	681	93.9
Subject 3	1999	1888	94.4
Subject 4	125	122	97.6
Subject 5	278	250	89.9
Subject 6	1581	1488	94.1
Subject 7	2656	2336	88.0
Subject 8	698	623	89.3
Subject 9	3355	3246	96.8
Subject 10	1033	965	93.4
Total	13727	12783	93.1

Fig. 11.13 Ratios of category numbers 1–18 and "Other"; 44.4% of the sentences were manually classified as category numbers 1–18

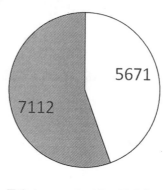

□ Category 1--18   ■ Other

Coincident sentences account for half of the total sentences in almost every category. We analyzed the kinds of differences that appeared in each category, because they might be key to finding a distance between categories. For example, "Meal" and "Cooking" are very close categories, because they include similar actions or behaviors. Table 11.9 shows the coincidence ratio for each category. The category of "Society" shows the worst score of all of the categories due to fewer sentences. It is possible that "Society" has different meanings for different people, so there are only a few characteristic words in this category, or that the meaning of "Society" overlaps heavily with other categories. On the other hand, the coincidence rates for "Go-out" and "Music" are around 60%. These categories might include characteristic words that express the category and that are far from other categories. These points have been analyzed closely.

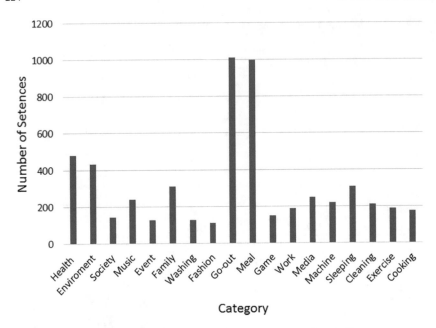

Fig. 11.14 Number of sentences manually classified as category numbers 1–18

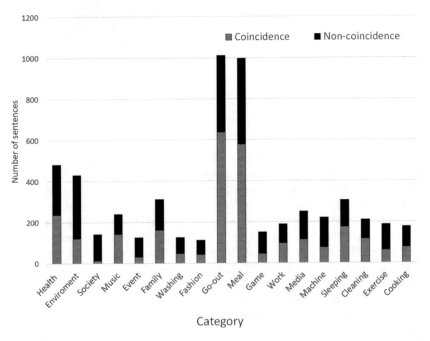

Fig. 11.15 Number of coincident and non-coincident sentences categorized by robot and manually

Table 11.9 Coincidence ratios for categorization by robot and manually

	Ratio [%]
Health	49.3
Environment	27.8
Society	9.2
Music	59.4
Event	25.4
Family	51.6
Washing	38.1
Fashion	37.5
Go-out	63.0
Meal	57.9
Game	29.8
Work	51.3
Media	45.4
Machine	33.9
Sleeping	57.2
Cleaning	56.2
Exercise	33.5
Cooking	43.5
Other	42.9

11.2.4 Summary

In this paper, we proposed a simple category estimation method for a chat robot. When chatting with speech recognition/synthesis, quick response is important. This method thus uses word matching with a keyword list and log history. Category estimation is difficult work even for humans. Thus, 10 students annotated the subjects' sentences manually, each sentence by 3 annotators. First, we compared the results of the three annotators. As the result, around 90% of the sentences could be categorized (2 or 3 persons annotated the sentence with the same category). However, over 50% of the sentences were categorized as "Other." We must consider the validity of this result. Moreover, we compare the results of manual categorization with that by the robot using the proposed method. In this paper, we focused on sentences that were categorized as NOT "Other" by humans. The results of the comparison show that around 50% of the sentences were categorized the same as by the human annotators. In future research, we will investigate the causes of the differences between robot categorization methods and human annotators. Moreover, we must examine how many and which kinds of categories a chat robot needs, for which it might be useful to compare the results of previous experiments and this experiment.

11.3   Collecting Common Sence from Interaction [1]

In recent years, aging has progressed in various countries around the world; the aging of the se population is particularly serious. A major problem of elderly people living alone is that the frequency of communication at home is low and the opportunity for contact with society is decreasing. Decreases in conversation can lead to depression. Therefore, in an earlier study, we conducted an experiment involving a chat robot for elderly people. However, from the results of the experiment, it can be said that a chat robot cannot as yet hold adequate conversations with the elderly. One reason is that robots make utterances that do not match the current season. Japan has four seasons, and seasonal information greatly influences conversation. Currently, the chat robot does not have the concept of seasons, so it makes utterances that do not match the current season, causing the user to lose the motivation to interact with the robot. Therefore, in this study, we proposed a method whereby the robot acquires the correct seasonal information by making a recovery if its utterance does not match the season. The user's response to the robot's utterance that does not match the season and the user's response to the robot's re-asking were characterized, allowing us to acquire correct seasonal information from users in the dialogue. In addition, by implementing the proposed method, it was possible to acquire data in interaction with the robot.

11.3.1   Research Background

In recent years, aging has advanced in various countries around the world. Aging in Japan is particularly serious. The number of elderly people living alone is increasing, which raises the problem of a low frequency of communication at home and reduced opportunities for contact with society. Because decreases in conversation can lead to depression, it is considered that regular conversation is important to counteract such a tendency. Therefore, in a prior study, Shimokawara et al. [18, 19] conducted experiments involving a chat robot for elderly people. In addition, in the study by Hirata et al. [6], in order to increase the robot's response sentences, they proposed a filtering method for extracting reusable sentences from the user's utterance to shape the robot's response sentences. However, a chat robot cannot as yet manage an adequate conversation with the elderly. One reason is that the robot makes utterances that do not match the season, as shown in Fig. 11.16.

In Japan there are four seasons, spring, summer, autumn, and winter, in a year. In spring, people enjoy cherry-blossom viewing. In summer temperatures exceed 35° and people go swimming in the sea and pools. In autumn, the mountains are bright red. In winter, snow falls and we warm our houses with a *kotatsu*. In this way, we enjoy the culture, events, and food associated with each season throughout a year, and we consider seasonal information as a matter of course in conversation. However,

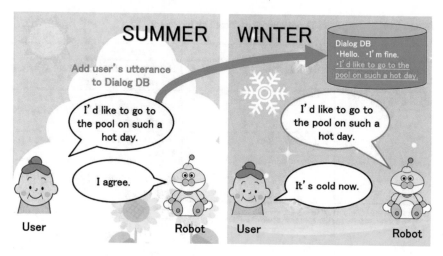

Fig. 11.16 The example of seasons in a chat robot

since the chat robot does not have a concept of the seasons, it can make utterances that do not match the current season when conversing with a human being.

11.3.2 An Example of Seasons in a Chat Robot

The utterances of the robot are divided into sentences that are restricted by the season and those that are not. "Tobacco is really bad for your health" is a sentence that can be uttered at any time of the year, so it is a "sentence not restricted by the season." Contrariwise, "The hot summer has continued; why don't we go swimming?" is a sentence supposed to be uttered in the summer, so it is a "sentence restricted by the season." As Table 11.10 shows, as the robot made this utterance in February, it does not match the season. About 15% of the robot's response sentences were restricted by the season, which means that the robot's utterance was likely to not match the situation one out of every six times during a dialogue. If the robot makes these utterances, the flow of conversation is interrupted. We also know that if the robot made such an inappropriate utterance even once, the user would have negative feelings toward the robot and would lose the motivation to interact with it. For this reason, it is necessary for the robot to understand seasons and make an utterance suitable for that situation.

Table 11.10 Sentences restricted and not restricted by the season

	Timestamp	Robot utterance
Sentences restricted	2015/2/22 9:38	Hot summer has continued. Why do not we go swimming?
By the season		(暑い夏が続いてきましたね。泳ぎに行きませんか?)
	2014/12/9 13:23	On hot nights, I use a fan when I go to bed.
		(暑い夜は、寝るときに、扇風機を使います。)
Sentences	2015/2/23	Tobacco is really bad for your health.
Not	21:19	(タバコは本当に健康に悪いですよね)
Restricted by the	2014/12/1 14:39	I am not good at singing songs and I'm tone deaf.
Season		(歌を歌うのは大の苦手で、音痴なんです。)

11.3.3 Purpose of This Study

Hirata et al. [6] manually added tags indicating whether robot's utterance was restricted by the season in order that the robot would avoid making an utterance that does not match the season in order to automate the judgment of sentences restricted by the season. Even in these studies in which teacher data is created and evaluated, an accuracy of 100% is difficult because the evaluation of the tagging reflects the subjectivity of the individual, and the criterion for this tagging varies from person to person. Also, it is currently not clear for which seasons a sentence is suitable for utterance. Seasonal judgment and recognition thus inescapably reflect individual subjectivity and depend on the place where the individual lives, so the robot needs to speak according to the user's seasonal feeling. Therefore, this study was conducted to reduce the utterances of a chat robot that do not match the season by adding tags for the time and season suitable for each user to the robot's response sentences.

11.4 Proposed Method

In order to add a seasonal tag for each user, we considered the problem of how the user can judge the time and season in dialogue with the robot. The user judges the time and

Table 11.11 Keywords associated with a given season

	Well-Known	Not Well-Known
Past	海 (beach)	ブルーベリー (blueberry)
Current (October)	紅葉 (autumn leaves)	(さんま) pike
Future	クリスマス (Christmas)	ぶり (yellowtail)

season, and we update the seasonal tag in real time according to the acquired seasonal information. This could facilitate communication suitable for the user because the robot's utterances are matched to those of each user. It should then be possible for the robot to avoid making an utterance that does not match the season. Therefore, in this study, we propose a method whereby the robot follows the user contrary to its own erroneous utterance if it makes an utterance that does not match the season and strikes the user as wrong. We analyze the characteristics of the user's response to the robot's utterances that do not match the season to facilitate the robot acquiring correct seasonal information from the user by re-asking the user. Therefore, we confirmed through experiments whether the robot could acquire seasonal information from the user and examined the kinds of features in the acquired information. We also confirmed which kinds of robot's utterances the user feels unsuitable, whether the feature is seen in the user's response at that time, and whether automated tagging is possible. We prepared [Keywords] associated with the season, [Template Sentences] to make an utterance that does not match the season, and [Recovery Sentences] to ask for correct seasonal information from the user.

11.4.1 Keywords

We prepared six Keywords, as shown in the Table 11.11, that are associated with seasons, comprising three types of words for a season that is suitable on, before, or after the experiment day. For each of the three types of words, there are two words: one well-known for the season (everyone commonly recognizes it) and one not well-known (people vary in recognizing it). Using Keywords that are suitable for the current season will allow us to identify their differences from Keywords not suitable for the current season. Having prepared words well-known for a given season and those that are not, we check whether there is a difference in users' responses to the two.

Table 11.12 Template sentences

Category	Template sentence	Example
Event	この時期は[*]で賑わっていますね。 (This time the city is crowded for [*].)	Christmas
	そろそろ[*]ですね。 (It is [*] soon.)	Halloween
	[*]が美味しい季節ですね. ([*] is delicious in this season.)	Yellowtail
Food	[*]が食べたくなる季節ですね. (It is the season when I want to eat [*].)	Blueberries
	そろそろ[*]に行きたくなる季節ですね。 (It is the season when I want to go to [*].)	The beach
Place	この時期は[*]が混んでいますね。 ([*] is crowded this season.)	The pool

11.4.2 Template Sentences

For utterances that do not match the season, we prepare Template Sentences as shown in Table 11.12. Template Sentences were prepared for each category of Keywords uttered by the robot; thus, by inserting Keywords not suitable for the current season, the robot can make an utterance that does not match the current season. The robot side utters this Template Sentences during the dialogue, and we examine the features of the user's response.

11.4.3 Recovery Sentences

If the user points out an erroneous season, the robot utters a Recovery Sentence shown in the Table 11.13 to re-ask the user. The robot utters the Keywords just like Template Sentences. We check whether correct seasonal information has been acquired from the user and examine the features of the user's response, as well as any differences in the user's response by the category of Keywords.

11.5 Experiment

11.5.1 The Process of the Experiment

The flow of the experiment is intended to promote dialogue in three steps: First is making an ordinary chat, second is making an utterance that does not match the season, and third is acquiring seasonal information from the user. As shown in Fig.

Table 11.13 Recovery sentences

Category	Recovery sentence	Example
Event	[*]はいつだったっけ?	Christmas
	(When is [*] ?)	
	[*]は何月だっけ?	Halloween
	(What month is [*] ?)	
	[*]の季節はいつだったっけ?	Yellowtail
Food	(When is the season of [*]?)	
	[*]の旬はいつだったっけ?	Blueberry
	(In which season is [*] delicious?)	
	[*]にはいつ行きますか?	Beach
Place	(When will you go to [*] ?)	
	[*]が賑わっているのはいつですか?	Pool
	(When is the [*] crowded ?)	

Table 11.14 Condition of experiment

Experiment date	2017/10/12-20
Number of subjects	6
Experiment time	30 min
Number of data points	208 (35 terms/person)
Topic	free
Experimental equipment	Windows10 PC ×2

11.17, while conducting a dialogue, the robot side (experimenter) intentionally makes an utterance including a "Keyword" that does not match the season and confirms the reaction of the user (subject) immediately afterwards. If a response is received from the user, the robot utters a Recovery Sentence to acquire seasonal information. The condition of experiment is shown in Table 11.14. The dialogue is done using the Wizard of Oz method of text chatting, and the robot side's utterance is done by humans. We tell the subject that they will have a dialogue with a robot. Regarding topics, we assume a chat robot, so they have a daily conversation. There were six subjects, all Information Systems students. None of the subjects had trouble typing, so almost real-time dialogue is possible. The dialogue time was 30 min per person, in which dialogues were conducted using the six Keywords. As a result, we acquired dialogue log data for 35 terms per person, totaling 208 cases.

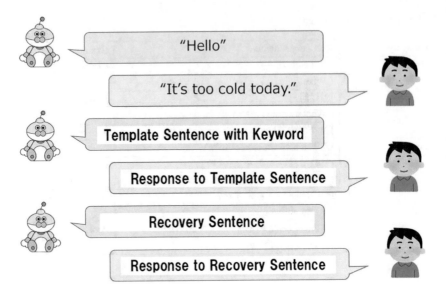

Fig. 11.17 The flow of the experiment

11.5.2 Instruction to Subjects

We gave two instructions to the subjects. First, the utterances were alternated and continuous transmission was prohibited because it was text chat, as continuous utterances would make it ambiguous which utterance answered which. This was intended to clarify the direct cause of human and robot utterances. Second, the user was to complete a questionnaire after the experiment. The subjects answered whether the robot's utterances felt unnatural or odd during the dialogue and why. This made it possible to determine what kind of robot's utterances made the user feel puzzled and why.

11.6 Results and Discussion

11.6.1 Users' Responses to the Template Sentences

The user's responses to the Template Sentences are shown in Tables 11.15 and 11.16. Gray indicates when the user pointed out an erroneous season to the robot; red indicates when the user felt something was wrong because of mismatch with the season; and blue indicates when the user felt something was wrong because of a sudden change of topic.

There are two major points to be made from the results. First, a comparison of the Keywords well-known for the season and Keywords that are not shows that in a Template Sentence, the former arouses greater discomfort than the latter. When feeling something wrong about a mismatch of the season, most users point it out to the robot. According to the questionnaire, when subjects did not point it out to the robot, they accepted the utterances because "it's a robot." "Feeling uncomfortable with the robot's utterance" and "Pointing it out to the robot" are not necessarily equal. Second, even to the same Template Sentence, some people pointed out the seasonal mismatch and others did not. Also, some subjects felt uncomfortable and others did not, depending on individual differences in the recognition and whether they knew the season or not. Many users felt something wrong due to a sudden change of topic, not seasonal mismatch.

The users' responses to utterances containing "beach" and "watermelon" whose season was before the experiment day, words such as "not" were seen, while the users' responses to "Christmas," whose season was after the experiment date, words such as "ahead," "yet," and "early" were seen. The presence of these keywords in the users' replies show that the users were pointing to the Template Sentence, suggesting that the robot should utter an appropriate Recovery Sentence.

11.6.2 Users' Responses to the Recovery Sentences

If the user pointed out the Template Sentence (gray cells in Table 11.15), the robot uttered a Recovery Sentence. The user's utterances in response to the Recovery Sentences are shown in the Table 11.17. By uttering the Recovery sentences, we were able to acquire seasonal information from the user. There are two major conclusions to draw from the results. First, users differ in their seasonal recognition of Keywords. One user answered "July" while another answered "August" to the same Keyword "beach." This shows that there were some seasonal deviations depending on the user. Also, one user answered "summer." Second, the characteristics of the users' replies differ by the category of the Keyword. Seasonal information was acquired in various forms, such as the month, the date, and the season. If the Keyword was an Event like "Christmas," we acquired a specific date, "December 25." If the Keyword was a Food like "watermelon" or "pike," we acquired the season, such as "summer" or "winter." If the Keyword was a Place like "beach" or "pool," we acquired the season, such as "summer," as well as the month, such as "July" or "August." From the above, it seems that the form of the appropriate seasonal information depends on the category of the Keyword. Thus, the form of the seasonal information can change for a given Recovery Sentence regardless of the category of the Keyword (for example, if the robot utters "Which month?," the user will reply with the month, whereas if the robot utters "Which season?," the user will reply with the season).

Table 11.15 Users' responses to the template sentences: well-known

	Keywords well-known for the season			
	Past (Before October)		Current(October)	Future(After October)
	Beach	Watermelon	Autumn leaves	Christmas
user01	なぜですか? (why?)		今が一番人多いでしょうね. (There are a lot of people the most now.)	まだ早いと思います. (I think that it is still early.)
user02	寒いです. 海には行きました. (It is cold. I went to the beach.)		いいこと言うじゃん. (It's a good thing to say.)	いや, さすがにまだ早い. (No, it's still too early.)
user03	今はそんなにいないんじゃないですかね. (I think there are not so many people right now.)		食べ物は好きなんですけど紅葉狩りはあまりやったことがないですね. (I like food but I have never gone to see autumn leaves.)	クリスマスはもうちょっと先じゃないかな. (Christmas is a while ahead, isn't it ?)
user04	そうですね. 魚が賑わっています. (I agree. Fish is crowded.)	スイカは夏ではないですか? (The season for watermelon is summer, isn't it ?)	紅葉狩りもいいですね。奥多摩の紅葉がきれいでした. (It is also nice to have autumn leaves. The autumn leaves of Okutama were beautiful.)	そうですね. 予定はもう決まっているのですか? (I agree. Is your schedule already decided?)
user05	入水者はそんなにかな. (There were few people in the sea.)	スイカはもう見ない頃よね. (I haven't seen a watermelon recently.)	だろうねえ. (I think so too.)	クリスマスになると少し早いんちゃう? (It is a little early to talk about Christmas.)
user06	最近は寒くてそんな気にはなりませんよ. (It has been cold recently so I don't feel that way.)	もうそんな季節じゃないですよ. (It is not that kind of season anymore.)	もう紅葉も散るころじゃないですか. (It is about the time autumn leaves are falling now.)	まだハロウィンも終わってないですよ. (Halloween has not ended yet.)

11.6.3 Additional Experiment

The experimental results showed that seasonal information can be acquired from users by uttering a Recovery Sentence if the user feels there is something wrong in utterances that do not match the season. We thus conducted an additional experiment, Experiment II, to verify whether the same result holds even for sentences actually uttered by the chat robot. Among the chat robot's actual utterance, we tried to utter sentences that 6 out of 7 people judged as restricted by the season as Template Sentences II. Therefore, as in the previous experiment, we confirmed whether users felt uncomfortable with the robot's utterances. If users feel that something is wrong, the robot can utter the Recovery Sentence in the same way. The Template Sentences II and condition of the experiment are shown in Tables 11.18 and 11.19.

Table 11.16 Users' responses to the template sentences: little-known

	Keywords little-known for the season		
	Past (Before October)	Current(October)	Future(After October)
	blueberry	pike	yellowtail
user01	ブルーベリー好きですか？ (Do you like blueberries?)	ブリもいいですけどやはりこの時期なら秋刀魚がおすすめです. (Yellowtail is delicious but pike is better; I recommend pike this season.)	ブリもいいですけどやはりこの時期なら秋刀魚がおすすめです (Yellowtail is delicious but pike I recommend this season.)
user02	ブルーベリー！目にいい！ (Blueberries! Blueberries are good for vision.)	秋刀魚の収穫量が減っているらしいよ. (The pike catch is decreasing.)	そうなんか、魚食べてねぇなあ最近. (I see. I have not eaten fish recently.)
user03	リンゴとぶどうもおいしいですよ. (Apples and grapes are delicious too.)	秋刀魚は好きですね. (I like a pike.)	ブリって11月なんですか？ (The season for yellowtail is November?)
user04		いいですね. でも秋刀魚は骨が多いですよね. (Pike's so good, butit's very bony.)	ぶりおいしいですよね. (Yellowtail is delicious.)
user05		おいしいよね (It is delicious.)	ブリって夏じゃなさそう. (The season for yellowtail Summer isn't really the season for yellowtail.)
user06		旬ですもんね (It is season.)	捕った魚を食べるんですか？ (Do you eat the fish you catch?)

Results and Discussion

The users' response when the robot uttered the Template Sentences II are shown in Table 11.20. The results indicate two major points. First, users often do not feel uncomfortable with the utterance of the robot, even if it is a sentence that 6 out of 7 people judged as being restricted by the season. The sentence "For the coming the rainy season, it is hard to do laundry," for example, is associated with the rainy season; however, regardless of the time it is uttered, it does not change the fact that "during the rainy season it is hard to wash clothes." As the result, many users think

Table 11.17 Users' responses to the Recovery Sentences

	Past		Current	Future	
	Beach (pool)	Watermelon	Autumn leaves	Christmas	Yellowtail
user01				12月25日です (It's December 25th.)	
user02	夏だよ (Summer.)			12月25日だよ (It's December 25th.)	
user03	海はやっぱり8月かいちばん人が多いと思いますよ (As for the sea, I think that most people are there in August)			クリスマスは12月です (Christmas is December 25th.)	寒ブリって聞くくらいだから、確かに冬な気がしますけどあまりよく知らないんですよね (It feels like winter, but I'm not certain.)
user04		夏だと思います (I think it is summer.)			
user05	7月とかならもう少しいるのかも (Maybe there are more people in July)	夏やな〜 (Summer.)		12月25日よ (It's December 25th.)	冬なのでは (I think it's winter.)
user06	夏ですね 7月、8月ころです (It's summer. About July and August.)	夏です (It's summer.)	9月下旬とかじゃないですか (I think it's in late September.)	クリスマスは12月25日です (Christmas is in December.)	

Table 11.18 Template sentences II

Robot's utterances that more than 6 out of 7 people associate with a season	
(1)	とりたてのキュウリは本当に美味しいですね (Fresh cucumber is really delicious.)
(2)	梅雨の時期は洗濯が乾かなくて大変ですね。 (During the rainy season, it is hard to wash clothes.)

Table 11.19 Conditions of experiment II

Experiment date	2017/12/12-15
Number of subjects	4
Experiment time	Up to 30 min
The number of data points	95 (24 terms/person?
Topic	Free
Experimental equipment	Windows10 PC ×2

Table 11.20 Users' responses (Experiment II)

	Robot's utterances	User's responses
(1)	とりたてのキュウリは本当に美味しいですね (Fresh cucumber is really delicious.)	どの食べ方がおすすめですか？ (How should I eat it?)
(1)'	この時期は,とりたてのキュウリは本当に美味しいですね (Fresh cucumber is really delicious **in this season.**)	それは夏ですね (It is summer.)
(2)	梅雨の時期は洗濯が乾かなくて大変ですね。 (During the rainy season it is hard to wash clothes.)	ですね (I agree.)
(2)'	これからの梅雨の時期は洗濯が乾かなくて大変ですね。 (**In the coming** rainy season, it will be hard to do laundry.)	これからは真冬ですよ (It will be midwinter from now on.)

that no matter in which season this sentence is uttered, there is no problem. Second, utterances for which the users felt something wrong include "this season" or "after this." Expressions such as "this time" or "after this" were thus targeted in Template Sentences II (as Table 11.20 (1)' (2)') Most users felt that the utterance did not match current season and pointed this out to the robot. These results suggest that the users do not feel something wrong in the seasonal topics of the robot, but do feel something wrong in such utterances as "this time" or "after this" because of the gap with the current season.

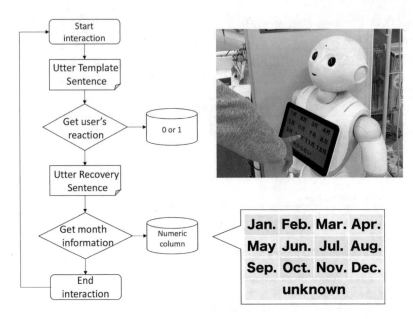

Fig. 11.18 System flow of interaction with pepper

11.6.4 Implementation of a System

We implemented this experiment in Pepper,[1] which allowed a variety of interactions and the acquisition of data through interfaces such as sensors and a touch panel. We created a system that acquires seasonal information from the touch panel through interactions with Pepper. The system flow is shown in Fig. 11.18. The Interaction begins as the user approaches. Pepper utters the Template Sentences with the Keyword as in the previous experiment. For each utterance, the users input through the touch panel whether they felt uncomfortable (uncomfortable: 1, not uncomfortable: 0). Pepper then utters the Recovery Sentences. The user could choose a month via the Keyword between "January" and "December," or "Do not Know" through the touch panel. [Keywords], [whether the user feels uncomfortable], and [month information] can be acquired as data. The results of carrying out the experiment using this system are given in Table 11.21. The month information selected by the user is different even with the same Keyword. These results also clearly show that users varied in their seasonal recognition.

---

[1] www.softbank.jp/robot/.

Table 11.21 Data acquired through interactions

Timestamp	Keywords	Uncomfortable	Month
2018/6/13 15:36:31	Christmas	1	December
2018/6/13 15:37:48	Autumn leaves	1	October
2018/6/13 15:37:07	Blueberries	0	May
2018/6/13 16:44:18	Autumn leaves	1	November
2018/6/13 16:44:57	Beach	1	August
2018/6/13 16:45:40	Pike	1	October
2018/6/13 16:46:18	Pike	1	September
2018/6/13 17:51:22	Blueberries	0	Do not know

11.6.5 Conclusion

In this research, we proposed a method to acquire seasonal information directly from the user during a dialogue in order to add seasonal tags for each user to a robot's utterances. This study tested whether it is possible to acquire correct seasonal information from the user by re-asking the user if the user feels that there is something wrong with the robot's utterances regarding the current season. We use Keywords, Template Sentences, and Recovery Sentences. The users reported when they felt uncomfortable with the robot's utterance that did not match the current season. The user's responses contained negative words such as "not" and words expressing time such as "early" or "yet." We were able to acquire the time information from the user in the dialogue via the robot's re-asking. Seasonal information was acquired in various forms such as dates ("December 25th"), months ("August"), and seasons ("summer"). Its characteristics were divided according to the category of Keywords. For example, by applying natural language processing technology to the sentence "Watermelon is delicious in summer," tags are given as "watermelon: August, summer." We think that it is possible to automatically update the seasonal tag of the response sentence database in real time during dialogue with the user. In addition, we implemented the proposed method in Pepper. By using the Pepper interface on a tablet, we acquired seasonal information. This research suggests that it is possible to acquire correct seasonal information from the user in a dialogue by re-asking the user if the robot detects the user's discomfort. It is also possible to acquire seasonal information for each user even if the seasonal recognition differs depending on the place of residence or the sensitivity of the individual. In future research, seasonal tags will be added to the robot's response sentences, and a dialogue experiment will be performed to compare the results for the dialogue system with seasonal tags with those of a conventional dialogue system.

In this study, we used text chat to acquire features from verbal information. However, in human conversation, it is obvious that non-verbal information such as expressions, talking speed, and tone of voice are major factors. For example, the user's discomfort toward an inappropriate utterance of the robot can be read from the characteristics of the user's non-verbal information. It may be possible to ease the user's dissatisfaction by adjusting the intonation and tone of voice of the robot's utterance. To that end, future experiments will use non-verbal information features on real machines.

References

1. Aoyagi, S., Hirata, K., Sato-Shimokawara, E., et al .: A method to obtain seasonal information for smooth communication between human and chat robot. In: 2018 Joint 10th International Conference on Soft Computing and Intelligent Systems (SCIS) and 19th International Symposium on Advanced Intelligent Systems (ISIS), pp. 1121–1126. IEEE (2018)
2. Bao, L., Intille, S.S.: Activity recognition from user-annotated acceleration data. In: Ferscha, A., Mattern, F. (eds.) Pervasive Computing, pp. 1–17. Springer, Berlin, Heidelberg (2004)
3. Chen, K.Y., Bassett, David R.J.: The technology of accelerometry-based activity monitors: current and future. Med. Sci. Sports Exer. 37(11), S490–S500 (2005)
4. Fukui, R., Mori, T., Sato, T.: An electrostatic capacitive floor sensor system for human position monitoring in a living space. Adv. Robot. 26(10), 1127–1142 (2012). https://doi.org/10.1080/01691864.2012.686346
5. Gratton, L., Scott, A.J.: The 100-year Life: Living and Working in an Age of Longevity. Bloomsbury Publishing (2016)
6. Hirata, K., Shimokawara, E., Takatani, T., et al.: Filtering method for chat logs toward construction of chat robot. In: 2017 IEEE/SICE International Symposium on System Integration (SII), pp. 974–979 (2017). https://doi.org/10.1109/SII.2017.8279349
7. Ho, Y., Sekine, N., Sato-Shimokawara, E., et al.: Motion pattern recognition using case-based reasoning for information providing. SICE Ann. Conf. 2011, 1284–1289 (2011)
8. Ikeda, S., Tabata, K.: Estimation of disability-adjusted life years $dalys$ in japan using a simplified method. Iryo To Shakai 8(3), 83–99 (1998). https://doi.org/10.4091/iken1991.8.3_83
9. Ishiguro, S., Kawagishi, Y., Yihsin, H., et al.: Motion recognition using 3d accelerometer sensor network for mobility assistant robot. In: 2012 IEEE International Conference on Fuzzy Systems, pp. 1–8 (2012)
10. Jacquemard, T., Novitzky, P., O'Brolcháin, F., et al.: Challenges and opportunities of lifelog technologies: a literature review and critical analysis. Sci. Eng. Ethics 20(2), 379–409 (2014). https://doi.org/10.1007/s11948-013-9456-1
11. Lo, V.W., Green, P.: Development and evaluation of automotive speech interfaces: useful information from the human factors and the related literature. Int. J. Veh. Technol. 2013 (2013). https://doi.org/10.1155/2013/924170
12. Menken, M., Munsat, T., Toole, J.: The global burden of disease study: implications for neurology. Arch. Neurol. 57(3), 418–420 (2000)
13. Ouchi, K., Doi, M.: Living activity recognition using off-the-shelf sensors on mobile phones. Ann. des Telecommun./Ann. Telecommun. 67(7–8), 387–395 (2012)
14. özyurt, O., Köse, C.: Chat mining: automatically determination of chat conversations' topic in turkish text based chat mediums. Expert Syst. Appl. 37(12), 8705–8710 (2010). https://doi.org/10.1016/j.eswa.2010.06.053
15. Pires, I.M., Marques, G., Garcia, N.M., et al.: Pattern recognition techniques for the identification of activities of daily living using a mobile device accelerometer. Electronics 9(3), 509 (2020). https://doi.org/10.3390/electronics9030509. http://dx.doi.org/10.3390/electronics9030509
16. Ravi, N., Dandekar, N., Mysore, P., et al.: Activity recognition from accelerometer data. In: Proceedings of the National Conference on Artificial Intelligence, pp. 1541–1546 (2005)

17. Sato-Shimokawara, E., Nomura, S., Shinoda, Y., et al.: A cloud based chat robot using dialogue histories for elderly people. In: 2015 24th IEEE International Symposium on Robot and Human Interactive Communication (RO-MAN), pp. 206–210. IEEE (2015a)
18. Sato-Shimokawara, E., Nomura, S., Shinoda, Y., et al. A cloud based chat robot using dialogue histories for elderly people. In: 2015 24th IEEE International Symposium on Robot and Human Interactive Communication (RO-MAN), pp. 206–210. IEEE (2015b)
19. Sato-Shimokawara, E., Shinoda, Y., Lee, H., et al.: An analysis of dialogue histories; in case of the elderly with a chat robot. J. Robot. Soc. Jpn. **34**(5), 309–315 (2016). https://doi.org/10.7210/jrsj.34.309
20. Shinoda, Y., Nomura, S., Lee, H., et al.: A Dialogue Analysis of Elderly Person with a Chat Robot, pp. 28PM1–2–2 (2015)

Chapter 12
Simulation-Oriented Data Utilization to Analyze Human Behavior in Urban Traffic Systems

Hideki Fujii, Kazuki Abe, Hideaki Uchida, and Shinobu Yoshimura

Abstract In this chapter, the authors describe data obtainment and utilization from the viewpoint of combining it with precise social simulations. More Specifically, taking traffic simulations as an example, the quantification of traffic system value is described using data output from a traffic simulator. Then, using the same simulator, an approach for estimating actual human trip demands behind the traffic system is explained, and the value of data obtainment is discussed. Through these case studies, the authors show that combining data utilization with simulationss is an effective method for analyzing the behavior of people who constitute social systems and for supporting the design of those systems.

Keywords Traffic simulation · Multi-agent system · Forward and inverse analysis · Policy assessment · Emission quantification · Traffic demand estimation · Reducing uncertainty

12.1 Introduction

Road traffic is one of the essential systems that support the movement of humans and goods in modern society. However, it also is the source of a wide variety of problems such as traffic jams, traffic accidents, regional environmental problems, and global warming.

H. Fujii (✉) · S. Yoshimura
The University of Tokyo, 7-3-1 Hongo, Bunkyo-ku, Tokyo, Japan
e-mail: fujii@sys.t.u-tokyo.ac.jp

S. Yoshimura
e-mail: yoshi@sys.t.u-tokyo.ac.jp

K. Abe
Vector Research Institute, Inc., 3-8-12 Shibuya, Shibuya-ku, Tokyo, Japan
e-mail: abe@vri.co.jp

H. Uchida
Osaka University, 2-1 Yamada-Oka, Suita-shi, Osaka, Japan
e-mail: uchida@see.eng.osaka-u.ac.jp

Y. Ohsawa (ed.), *Living Beyond Data*, Intelligent Systems Reference Library 230,
https://doi.org/10.1007/978-3-031-11593-6_12

Various countermeasures (such as of vehicle performance improvements, traffic control by traffic lights, road network modifications, laws and regulations, the introduction of Intelligent Transport Systems (ITS) technology, and the promotion of public transportation) are being considered to minimize or resolve those problems.

Furthermore, since it is very difficult to restore a road environment to a previous condition once it has been changed, it is strongly desirable to accurately estimate the impact of transportation policies quantitatively. It is for that reason that simulations have been playing an important role in the field of traffic engineering, and why various types of traffic simulators have been developed and utilized [3, 16].

In this chapter, the authors first report on forward analysis using a multiagent-based traffic simulation to obtain precise data. Herein, a precise simulation that reflects reality well is called a virtual social experiment. Some kinds of data obtained from virtual social experiments have the potential to change human behavior patterns even though they cannot be observed in the real world. Later, some case studies will be explained.

In the latter half of this chapter, we will discuss methods that can be used to estimate a part of the simulation input data. As with other social simulation methods, parameters representing the internal or psychological states of an individual human being are difficult to obtain directly in the field of traffic simulation.

In particular, even though traffic demand is unobservable in a practical sense, it must be input into a simulator to create a realistic traffic simulation. To address this issue, approaches that estimate traffic demand from the link traffic volume, which is observable by traffic surveys, can be adopted. In this study, the use of inverse analysis combined with running traffic simulations iteratively will be described as an example, and optimization of the observation points to reduce the uncertainty of traffic demand estimates will be explained. We also provide an index that can be used

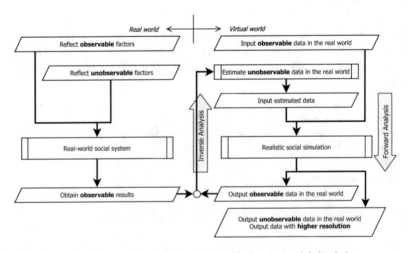

Fig. 12.1 Conceptual image of forward and inverse analysis using social simulation

to obtain new helpful data actively under resource constraints, while a conceptual image of forward and inverse analysis is shown in Fig. 12.1.

12.2 Quantifying Value of Traffic System by Forward Analysis

When traffic phenomena are regarded as a mixture of complex systems produced by numerous human beings who possess intelligence, goals, intentions, as well as individualities, a multiagent system is useful for describing microscopic car behavior. In this section, case studies performed using a multiagent-based traffic simulator, named ADVENTURE_Mates,[1] which the authors have been developing [13, 28], are described as examples of quantifying the value of traffic systems.

12.2.1 Overview of Simulation Model

12.2.1.1 Definition of Road Environment

A two-level road network model based on directed graphs is employed in this simulator and a sample network is shown in Fig. 12.2.[2] The virtual driving lane is the fundamental unit for modeling the actual road structure in the bottom layer network. The car agent maneuvers are restricted to movement along the lane, except when lane-shifting. Each lane is equipped with various kinds of information related to its length, connections with other lanes, speed limit, and other accompanying attributes. The road environment provides such information if it is requested by the agent.

Two types of lane bundle objects, basic road segments and intersections, are located in the top layer which represent a global roadmap. Each object consists of virtual driving lanes and their connectors. This layer is used for route searching.

12.2.1.2 Definition of Car Agent

A simple flowchart of the simulation, in which a time interval $\Delta t = 0.1$ [s] is used, is shown in Fig. 12.3. At each time step, car agents are generated according to the specified traffic demand at each terminal node at the road network. The agent searches for its route from the origin node to the destination node at the time it is generated. When the car agent arrives at the destination and completes its trip, it is removed

---

[1] The open-source version of MATES (Multi-Agent-based Traffic and Environment Simulator) released on the ADVENTURE project website (https://adventure.sys.t.u-tokyo.ac.jp/).

[2] Note that cars are driven on the left side of roads in Japan.

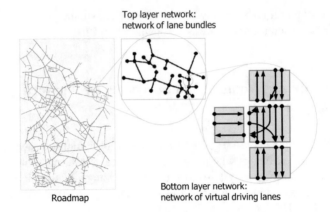

Fig. 12.2 Layered road network

from the simulation. A car agent always percepts its surroundings while driving. That information, which determines its acceleration, is also used to update its velocity and position.

When searching for the driving route, the A* algorithm [14] is implemented into the simulator. Here, it should be noted that the simulator does not include a planning process in which the origin and destination (OD) nodes are decided. Hence, users will need to input a proper OD matrix *a priori* (details will be provided later). The route that minimizes the trip distance or the expected trip time from the origin to the destination is selected by the A* algorithm. A search is conducted for the optimum route every time a car agent is generated at a terminal node, and the results are stored and reused. In the case of a search for a route that minimizes the expected trip time, the cost of each link is given as the required time average of all cars that have already passed the link, and this average is updated every 10 min.

After determining the global route on a road network, each car follows the selected route and drives from the origin to the destination. The simulator employs the generalized force model (GFM) [15] in order to determine the acceleration of a car agent. This model concept is shown below:

$$\frac{dv_i}{dt} = \frac{v_i^0 - v_i(t)}{\tau_i} + f(x_i(t), v_i(t), x_{i-1}(t), v_{i-1}(t)) + \xi_i(t), \qquad (12.1)$$

where $x_i(t)$, $v_i(t)$ is the position and speed of car agent i at time t, respectively. The first term on the right side represents the acceleration toward the car's desired speed v_i^0 with relaxation time τ_i. The second term (≤ 0) represents the virtual repulsive force from interactions with the leading car agent $i - 1$. The third term may be used to include individual variations of driver behavior.

The GFM is a model in which the acceleration of car is solely based on the distance and the speed difference from the leading car. However, since speed determinants should also include the traffic lights, the forward intersection situation, and other

Fig. 12.3 Overview of simulation flow using ADVENTURE_Mates

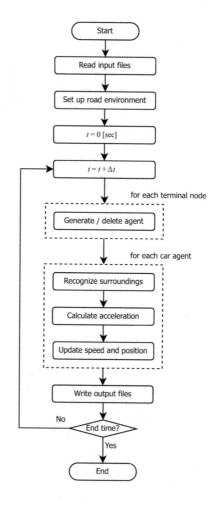

urban traffic conditions in addition to the leading car, the authors have expanded the GFM so that the virtual leading cars reflect the forward road conditions, and Eq. (12.1) is to be applied to each virtual leading car.

12.2.2   Utilization of Simulation Output Data

In the latter half of this chapter, selected applications of traffic simulation from the authors' previous work will be described.

Fig. 12.4 Simulation
screenshot showing the
intersection in front of the
station

12.2.2.1 Virtual Social Experiment on Tram Railway Extension

The first case is a typical analysis of traffic engineering. In Okayama City, which
is the capital of Japan's Okayama Prefecture, discussions were held regarding a
plan to extend the tramway into the station square in order to improve access and
convenience of movement for train users around the downtown areas. However,
because the tramway would cross the large intersection in front of the rail station,
and because the duration of green light time for car traffic would be reduced to
allow tram passages, private car users were concerned that the project would have a
negative impact on car traffic. With this point in mind, we attempted to quantitatively
clarify the impact of the proposed extension by conducting a traffic simulation [13].

Figure 12.4[3] shows a simulation screenshot around the intersection in front of the
Okayama Station. The red, black, blue, and orange rectangles represent private cars,
taxis, buses, and trams, respectively, while the green dots represent pedestrians. In
this simulation, pedestrian and tram models are included in addition to cars. The small
squares around the intersections indicate traffic light colors or passage permissions.

As an example of the simulation results, Fig. 12.5 (left) shows a position comparison of the rearmost stopped car agent in the first lane (where cars are permitted to
turn left or go straight) on the north-side road of the intersection in front of the station. The mean, minimum, and maximum values of the positions that were measured
once per signal cycle are indicated in the figure. "BASE-2" in Fig. 12.5 indicates
the result reflecting the current condition, while "BASE-1" indicates unrealistic conditions that ignore the existence of pedestrians. "EXT-32" and "EXT-64" are the
results that suppose the tramway extension, and the number represents the number
of seconds that the green light for cars will be shortened after the tramway extension.

---

[3] The background image was imported from OpenStreetMap (https://www.openstreetmap.org/).

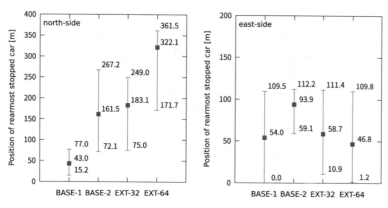

Fig. 12.5 Position of the rearmost stopped car agent (left: in the first lane of the north-side road, right: in the first lane of the east-side road)

EXT-32 is a realistic and reasonable scenario, while EXT-64 is a fictional one used only for comparison purposes. The cycle of the traffic light at the intersection is fixed to 150 seconds.

In the comparison between BASE-2 and EXT-32, even though the average value is declining slightly, it can be seen that the change is sufficiently small compared with the daily fluctuation level. Figure 12.5 (right) shows the position of the rearmost stopped car agent in the first lane (where cars are only permitted to turn left) of the east-side road. In EXT-32 and EXT-64, the congestion length is seen to decrease compared with BASE-2, and the traffic flow is found to become moderately smooth due to the changes in the signal control patterns that accompanied the tramway extension.

The results of the simulation facilitated discussions in the local community, including the local government, prefectural police, transportation operators, the local merchants' association, and residents, which resulted in the decision to implement the extension plan. This means that the data obtained from the simulation supported the people's decision-making.

12.2.2.2   Evaluation of Vehicle CO$_2$ Emissions

The simulation can output not only real-world observable quantities such as traffic volume and congestion length, but also the position, speed, and acceleration of each vehicle at each step. By combining the high-resolution vehicle behavior data obtained via the simulation with the exhaust emission database obtained by actual car experiments, it was possible to obtain the total amount of carbon dioxide (CO$_2$) emissions caused by the whole car traffic system in a particular region, as well as the emission history of each vehicle based on the driving conditions [11].

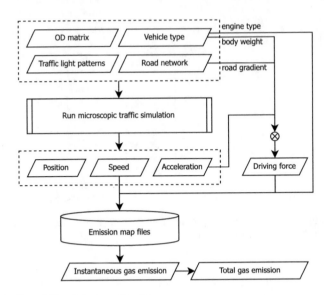

Fig. 12.6 Outline of CO_2 emission estimation

This approach, which is based on a microscopic traffic simulation in this study, can provide a way to enhance data resolution. In most conventional estimation methods, on the other hand, emissions are calculated by multiplying the trip length and/or gasoline consumption by specific emission factors. However, such methods ignore detailed car behaviors that reflect interactions with other cars and the effects of traffic lights.

The Japan Clean Air Program 2 (JCAP2) emission map files published by Japan Petroleum Energy Center (JPEC)[4] are utilized as a vehicle emission database in this study. This database indicates the measured correspondence between car driving data (speed, driving force, vehicle type, and engine type) and the amounts of exhaust gases, such as carbon monoxide (CO), CO_2, hydrocarbon (HC), nitrogen oxide (NOx), and suspended particulate matter (SPM).

The estimation flow outline is shown in Fig. 12.6. First, the traffic simulation is executed by inputting the road network data and the OD matrix. As a post process, the driving force is calculated from the acceleration data at each time step and the road gradient. The instantaneous amount of gas emissions of each vehicle is obtained by matching the obtained driving force with the speed history, vehicle type, and engine type contained in the simulation output data to the vehicle emission database. By adding up these instantaneous emissions spatiotemporally, the total amount of emissions per region or per vehicle can be determined.

The following is an example of CO_2 emission estimation in an urban road network. The simulation target is the northwest area in Kashiwa City, Chiba Prefecture, Japan. The road map is shown in Fig. 12.7. It has an east-west width of 9.2 km, a north-south

---

[4] https://www.pecj.or.jp/en/.

Fig. 12.7 Target area for vehicle CO_2 emission estimations

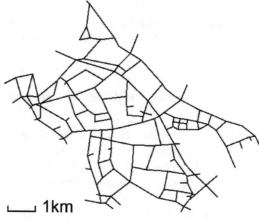

Fig. 12.8 Estimated CO_2 emissions

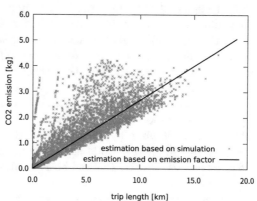

length of 7.3 km, and contains 172 nodes and 229 links. A one-hour simulation was run with 10,715 vehicles on this map.

The result is shown in Fig. 12.8. Each dot in the chart represents the relationship between the trip length and the CO_2 emissions of each car. The line in the chart represents the estimation result produced by the conventional method using an emission factor of 267 g-CO_2/(km·vehicle) [22].[5] In a complex network having numerous intersections and traffic lights, the emission amounts are not always proportional to trip length. The total amount of emissions in this area in one hour was 1.70×10^4 kg-CO_2 when calculated by the method proposed in this research, while 1.45×10^4 kg-CO_2 was calculated using the conventional method.

---

[5] This emission factor was derived from some statistics such as total vehicle miles traveled and gasoline consumption, which reflected the driving situations in the real world and included energy consumption by accessories. Consequently, the value is much higher than the certificated ones measured in the laboratory or on the test course.

Although this example was an environmental impact assessment, the method can also be used to estimate the economic value of road construction and maintenance projects in response to changes in the driving conditions of individual cars [12].

12.3 Estimating Unobservable Demand by Inverse Analysis

In order to simulate the real-world traffic phenomenon using the microscopic traffic simulator introduced in the previous section, it is necessary to input traffic demand into the simulator appropriately. Traffic demand is typically input as an OD matrix, which describes demands between origin and destination node pairs in a road network.

Since the OD matrix cannot be observed directly, it has to be estimated in some way. The approaches for OD matrix estimation can be roughly classified into two categories. The first one is based on population distribution. This approach is commonly used for traffic and civil planning using the four-step model [23]. Since the population distribution is derived from traffic census data, simulation users do not need to measure the actual traffic flow. However, the census data resolution is low, and the accuracy of the estimated results is not guaranteed.

The second approach is based on an inverse analysis of the link traffic volume data. The link traffic volume is the traffic volume counted at a fixed location. The OD matrix is optimized by minimizing the difference between the observed link traffic volume in the real world and the reproduced link traffic volume based on the estimated OD matrix. This can be accomplished with a bi-level programming approach [4].

Furthermore, the inverse analysis approaches can be classified into two categories depending on the method used to solve the direct problem involved in the estimation process. Here, the direct problem refers to the assignment from the OD matrix to the link traffic volume. The solution to the direct problem can be obtained analytically, or approximated with the equilibrium assignment algorithm [18].

Estimation of the OD matrix using the equilibrium assignment algorithm has been shown to be effective for large-scale road networks [20]. However, the results of the equilibrium algorithm may not be consistent with microscopic traffic models. The alternative is to use the traffic simulator to solve the direct problem in the estimation flow. In this section, an OD matrix estimation method using a microscopic traffic simulator and discussions about the accuracy and stability of the estimation [1] will be introduced.

12.3.1 Estimation Methodology

An outline of the adopted estimation method is shown in Fig. 12.9. The proposed method consists of the following steps:

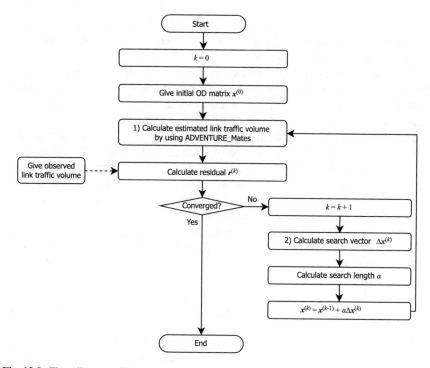

Fig. 12.9 Flow diagram of the proposed method for estimating the OD matrix

(1) Calculate link traffic volume from the OD matrix
(2) Update the optimal OD matrix solution.

ADVENTURE_Mates is used in the calculation step (1), and the Levenberg-Marquardt gradient method (LMM) [19, 21] is used in the update step (2). After calculating the link traffic volume, the link traffic volume residuals are evaluated. The optimal estimated OD matrix is updated using these residuals. These steps are then iterated until the estimated OD matrix converges to a tolerance level.

12.3.1.1 Formulation

The OD matrix is estimated so as to minimize the residual norm of the link traffic volume between the estimated and the observed values. This process can be expressed as follows:

$$F = \|r\|^2 \to \min., \tag{12.2}$$

where r is the residual vector link traffic volume defined as:

$$r = \bar{Q} - \hat{Q}(x), \tag{12.3}$$

x is the vector-formed OD matrix, $\bar{Q}$ is the observed link traffic volume, and $\hat{Q}(x)$ is the estimated link traffic volume. A traffic simulation is used to calculate $\hat{Q}$ link traffic volume from the OD matrix x.

Here, we define the dimensions of the vectors. N denotes the dimension of x, i.e. the number of OD pairs, and M is the dimension of Q, i.e., the number of observation points. Because, generally speaking, N is larger than M, more variables have to be estimated from smaller datasets in the OD matrix estimation.

In this research, the Euclidean norms are employed with the assumption of no variance and no covariance in the observed link traffic volume. If the variance and covariance of the observed link traffic volume are known, the residuals can be regularized with consideration of the variance and covariance by using the Mahalanobis distance to calculate the norms. However, this requires multiple traffic volume observations, which imposes significant observational costs.

For simplicity, we assume that the traffic flow is in a steady state and the relationship between the OD matrix and the link traffic volume is linear. The assumption of a linear relationship is appropriate if no traffic jams occur. Here, the linear relationship can be described below using a matrix J:

$$\hat{Q}(x) = Jx. \tag{12.4}$$

The assumption of a steady state means that both the OD matrix and the link traffic volume are stationary in time. This assumption is reasonable if an appropriate time period is used, such as a short time period where no drastic changes in traffic flow occur.

J in Eq. (12.4) corresponds to the Jacobian matrix of F with respect to x, which is an $M \times N$ matrix. Since this system has no explicit constitutive equation, J has to be approximated. The components of J are estimated as in Eq. (12.5) using the results of the traffic simulation.

$$J_{i,j} = \begin{cases} 1 & \text{(when the link } i \text{ is included in the route of OD pair } j) \\ 0 & \text{(otherwise)} \end{cases}, \tag{12.5}$$

Since the route choice behaviors of car agents may be changed according to changes in x, J is calculated at each iteration.

12.3.1.2 Solution Updating Step

The OD matrix is updated using one of the gradient methods. This updating step is given in the following general form:

$$x^{(k+1)} = x^{(k)} + \alpha \Delta x^{(k)} \tag{12.6}$$

where α is the search length coefficient, and Δx is the search direction vector. $\bullet^{(k)}$ means a variable for k-th iteration step. In this study, the LMM is used. The definition of Δx in the LMM is given as follows:

$$\Delta x = -H^{-1}G, \tag{12.7}$$

where G is the gradient of the objective function F with respect to x, i.e. $G = -J^T r$, and H is the Hessian matrix of F with respect to x.

The steepest descent method (SDM) is commonly used in optimization problems because it only requires a gradient matrix. However, the convergence is slower than other methods. On the other hand, the LMM is based on the Gauss-Newton method (GNM) and is applicable to nonlinear problems. The Hessian matrix is assumed to be the product of Jacobian matrices: $H = J^T J$ in these methods. However, the GNM can only be applied when $N < M$. In the LMM, the Hessian matrix is regularized as follows:

$$H = J^T J + \lambda D \tag{12.8}$$

where λ is a regularization parameter for LMM and D denotes a diagonal matrix, such as the identity matrix. The additional diagonal matrix makes the Hessian matrix non-singular so that its inverse can be calculated. Marquardt proposed setting λ to a large value and then decreasing it with successive iterations [21]. If λ becomes zero, Eq. (12.8) is equivalent to the updating scheme for the GNM. If λ is sufficiently large, the search direction approaches that of the SDM. According to Marquardt's proposal, the initial iteration steps have the robustness of SDM, and the later iteration ones have a high rate of GNM convergence. In this study, evaluating F is computationally expensive. Thus, λ is set to be decreased from an initial value λ_0 (> 0) at a constant rate γ ($0 < \gamma < 1$) with successive iterations as follows:

$$\lambda = \lambda_0 \gamma^k \tag{12.9}$$

The convergence is judged by comparing relative residual norm (RRN) $\|r\|/\|\bar{Q}\|$ and a tolerance ε. Since RRN is normalized by the observed link traffic volume $\bar{Q}$, it can be compared among different networks.

12.3.1.3 Applying Non-negative Constraint

A non-negative constraint of traffic volume is necessary for an OD estimation. When the OD matrix satisfies the constraint, the link traffic volume will always satisfy it as well. Thus, it is sufficient to apply the constraint solely to the OD matrix. Since the original LMM has no constraint for variables, the following two method types, which are used to introduce non-negative constraints into LMM, will be examined later.

The common assumption used in both of the methods is as follows. In this study, the non-negative constraints are applied stricter, i.e., $x_i > \delta$ is required for all i. δ is a small value but larger than zero. A small value of x_i reduces some estimated link traffic volume to zero. If this occurs, some components of the Jacobian matrix also become zero, and the zero components are then propagated to components of Δx. Consequently, the search direction becomes limited.

Based on this assumption, the first method (named "Method A") is designed to strictly satisfy the constraint. Here, the initial value of α is set to 1. If $x_i^k + \Delta x_i^k < \delta$ for any element i, α is modified to be somewhat smaller in order to satisfy $x_i^k + \Delta x_i^k = \delta$ for all elements i. Although it does not modify the search direction from that given in the LMM, the residuals will still be expected to stop decreasing when α approaches zero with successive iterations.

The second method (named "Method B") uses heuristics. Here, the search length coefficient α is fixed strictly to 1 strictly. After the solution updating step, the updated x_i is forcibly modified to be δ individually when $x_i < \delta$ for any index i. Although it creates a different search direction from that made in the LMM, since α does not become zero, the iteration is not expected to halt prior to the convergence.

12.3.2 Numerical Experiments

In order to assess the validity of the proposed estimation method, it is necessary to evaluate some of the indices that are discussed in terms of traffic engineering, such as the regression factor of the reproduced link traffic volumes to the observed ones and the correlation coefficient between these two link traffic volumes. Here, the proposed method will be applied to cases by considering artificially-generated link traffic volume. Using these results, the adequate tolerance ε and the proper method used to consider the non-negative constraint, as well as the accuracy and stability of the estimation method, will be discussed.

12.3.2.1 Simulation Conditions for Experiments

The map used in the numerical experiments is a section of an actual road network in Tokyo (shown in Fig. 12.10). Within this 3 km × 3 km area, the number of OD nodes $n = 26$, the number of OD pairs $N = 600$, and the number of links where the traffic volume is observed $M = 22$. Because it includes one-way links, the number of OD pairs N is less than $n(n - 1)$. In order to update the regularization parameter λ in each step, $\lambda_0 = 10$ and $\gamma = 0.25$ are used (Eq. (12.9)).

Next, we prepare the observed link traffic volumes as references. In this study, the simulated datasets are used for all experiments. We ran the traffic simulator with OD matrices that satisfy the assumption of linearity described in Sect. 12.3.1.1 and the measured link traffic volumes. The condition was $x_i < 14, \forall i$ for the experiment. As far as the datasets that have been generated under the conditions are concerned, the existence of the solution is guaranteed. Additionally, other cases in which artificial noise has been introduced are considered in order to examine the accuracy and

Fig. 12.10 OD matrix
estimation target area

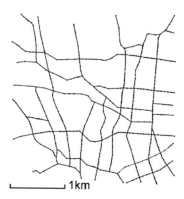

1 km

stability of the proposed method. Noise is added to the observed link traffic volume
$\bar{Q}$ as follows:

$$\bar{Q}_{\text{noisy}} = \bar{Q} + \delta \tag{12.10}$$

Here, δ is a noise vector, where each element δ_i follows a uniform distribution
whose upper and lower limits are within $\pm 10\%$ of $\bar{Q}_i$, For these experiments, 10
kinds of observed link traffic volume are prepared using 10 different random seeds.

12.3.2.2   Estimation Results

The proposed estimation method minimizes the RRN. This is important because a
smaller RRN means that the link traffic is more reproducible, which satisfies the
necessary condition for an appropriate OD matrix estimation. The RRN transitions
after estimation in the cases with and without noise are shown in Figs. 12.11 and
12.12, respectively.

In this experiment, the RRN transitions in both cases are similar. In several early
iteration steps, the RRNs in Method B are smaller than those in Method A. However,
after that, RRNs in Method B become larger than those in Method A. If the tolerance

Fig. 12.11 RRN transition
in estimation with clean data

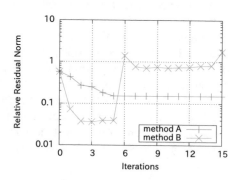

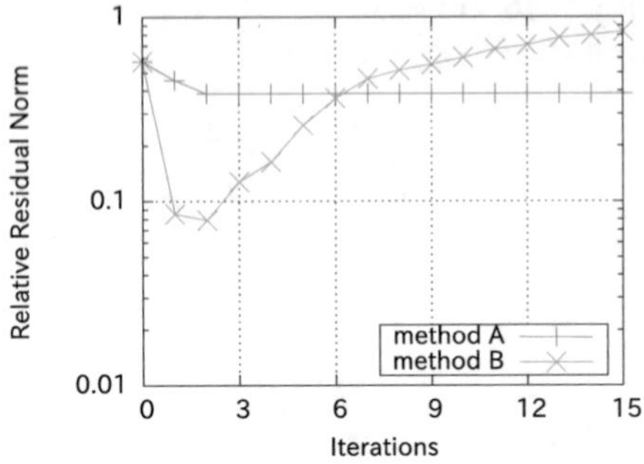

Fig. 12.12 RRN transition
in estimation with noisy data

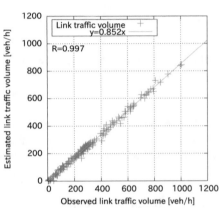

Fig. 12.13 Link traffic
volume reproduced from
clean data using Method A

for convergence ε is set to 0.08, only Method B converges and the optimal solution
is obtained.

Here, the reproduced link traffic volumes scoring minimum RRN in each case:

- using Method A with clean data (data without noise),
- using Method B with clean data,
- using Method A with noisy data, and
- using Method B with noisy data.

The results of each case are shown in Figs. 12.13, 12.14, 12.15, 12.16. The x-axis
is the observed link traffic volume $\bar{Q}$, while the y-axis is the link traffic volume
$\hat{Q}(x)$ reproduced from the estimated OD matrix x. These link traffic volume figures
indicate the following two evaluation indices: the regression factor a (the gradient
of the regression line $y = ax$) and the correlation coefficient R. If the OD matrix is
estimated accurately, both indices are close to 1.

These results show that the regression factor a is distant from 1 when Method A is
applied, i.e., 0.852 without noise and 0.618 with noise. In contrast, when Method B
is used, the difference from 1 is at most 0.021. The correlation coefficient R almost
equals 1 for all constraint methods and cases. These results suggest Method B is more

Fig. 12.14 Link traffic volume reproduced from clean data using Method B

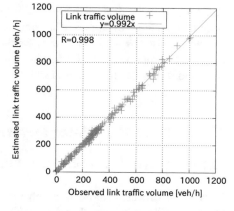

Fig. 12.15 Link traffic volume reproduced from noisy data using Method A

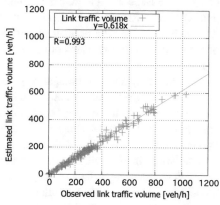

Fig. 12.16 Link traffic volume reproduced from noisy data using Method B

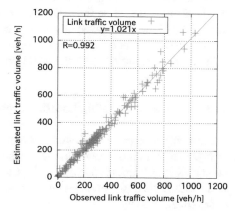

accurate and stable than Method A. Additionally, these accuracy levels are equal to or better than those of studies using similar methods [26].

The proposed method can be used for the estimation of the people's trip requirements hidden in observable traffic phenomena in the way described above. Precise simulations play key roles here as well.

12.3.3   Value of Obtaining Data and Reducing Uncertainty

12.3.3.1   Uncertainty in OD Matrix Estimation

In the OD matrix estimation, the number of variables to be optimized (the number of OD pairs) N is much larger than the number of constraints (the number of observation points) M. In other words, we have to deal with the inverse problem of ill conditions. Although increasing the number of observation points is considered to be a means of reducing uncertainty, it is not practical (in terms of cost) to observe all the link traffic.

Uncertainty quantification is an approach that aims to quantitatively assess and, if possible, reduce the uncertainty of a model. Uncertainty is generally categorized into "aleatory uncertainty" and "epistemic uncertainty" based on its causes. Aleatory uncertainty is the scatter caused by the inherent differences and variations in the system under consideration, which cannot be reduced by accurate modeling and data accumulation. In contrast, epistemic uncertainty is the scatter and errors caused by incomplete modeling or lack of knowledge, which can be reduced by more accurate modeling and better data accumulation.

In the estimation of the OD matrix, the effect of including probabilistic factors in the distribution model on the estimation results can be regarded as aleatory uncertainty. The difference between the estimated and the true values due to incomplete models, over or under observation points, and the observation errors in the link traffic volume can be regarded as epistemic uncertainty.

12.3.3.2   Traffic Counting Location Problem

Traffic congestion points and entrances to major roads have been empirically chosen as observation points. However, there is still room for optimizing the observation point locations. This is called the Traffic Counting Location Problem (TCLP) [6, 7]. In the TCLP, the observation points are basically located at the links, rather than the nodes (intersections).

Algorithms used to search for optimum observation locations have been studied previously. For example, Lam et al. [17] proposed a method that preferentially selects points with heavy traffic, and Yang et al. [25] proposed a method that selects points to maximize the number of captured OD pairs. Separately, Yim et al. [27] proposed a method that maximizes the total traffic of the captured OD pairs, which will be

Table 12.1 Example of captured OD pairs

	Observation link	Captured OD pairs	The number of captured OD pairs
Example 1	l_a	$O_1 \rightarrow D_1$	2
		$O_1 \rightarrow D_2$	
Example 2	l_c	$O_2 \rightarrow D_1$	1
Example 3	l_e	$O_1 \rightarrow D_1$	2
		$O_2 \rightarrow D_1$	
		$O_1 \rightarrow D_1$	
Example 4	l_a & l_c	$O_1 \rightarrow D_2$	3
		$O_2 \rightarrow D_1$	
		$O_1 \rightarrow D_1$	
Example 5	l_a & l_c	$O_1 \rightarrow D_2$	3
		$O_2 \rightarrow D_1$	

Fig. 12.17 Simple network used to capture OD pairs

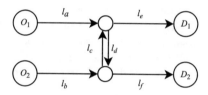

described below. Chootinan et al. [6] proposed two optimization methods; the one is a method that minimizes the number of observation points and the other is a method that maximizes the number of captured OD pairs. Fei et al. [9, 10] proposed an optimization method that takes into account the error in the observed traffic.

These studies primarily discussed the method used to place all the observation points in the target area at one time. However, in reality, we often need to add additional observation points to complement the existing data in order to improve the estimation accuracy. For example, Ehlert et al. [8] proposed a method to place observation points using information such as traffic distribution rates estimated from existing observation data, while, Wang et al. [24]. proposed a method that assumes errors in the estimation results from existing data, and then places new observation points to minimize them. Separately, Chen et al. [5] showed that the method of increasing the number of captured OD pairs after placing them on major roads was more effective than methods that only consider the number of captured OD pairs.

A captured OD pair is defined as the inclusion of an observation point in the path of that OD pair. As an example, Table 12.1 shows some relationships between observation links and captured OD pairs in the network shown in Fig. 12.17. In this example, there are two origin nodes: O_1 and O_2, two destination nodes: D_1 and D_2, and four OD pairs ($2 \times 2 = 4$). In addition, there are six directed links l_a, l_b, l_c, l_d, l_e, and l_f. The existing methods, such as the one proposed by Yang et al., have noted that the placement of observation points can be optimized by maximizing the number of captured OD pairs. According to their proposal, in cases where one observation point is placed (Examples 1–3 in Table 12.1), it is preferable to place it on link l_a or l_e rather than to place it on link l_c due to the increasing number of captured OD pairs.

In contrast, as mentioned above, there is uncertainty in the estimation method. Even if information on important links can be obtained, it is not always possible to reflect it accurately within the estimation. Therefore, it is necessary to take into account not only the importance of the link but also the difficulty of the estimation. In the case using the estimation method described in this section, a link with a higher number of captured OD pairs is likely to reflect a more complex traffic situation. Therefore, it is difficult to determine the traffic demand of the OD pairs that it captures. In other words, such links can possess a high degree of epistemic uncertainty. In response, the authors regarded the number of captured OD pairs as the size of the search space and proposed a method to maximize the importance of the links given by parameters such as traffic volume, while simultaneously limiting the search space [2].

Although the indices used for the TCLP are different from each other, all of these studies attempt to value the act of obtaining data from existing knowledge and data. Additionally, iterative approaches using precise simulations can also help these considerations, since it is practically impossible to conduct examinations on huge, complex, and dynamic traffic systems in the real world.

12.4 Conclusion

In this chapter, using a traffic system as an example, the authors introduced two approaches to utilizing data and simulations together.

The first approach is the forward simulation, which uses precise simulations to generate higher resolution data or data that cannot be measured in reality. An assessment of the change in the length of traffic jams with the implementation of the tramway extension project and an estimation of the environmental impact of CO_2 resulting from car traffic were described in Sect. 12.2. The data generated in this approach can support people's decision-making and sometimes promote beneficial behavioral change.

The second approach works it in the opposite manner of the first one. The output from the simulation and partially-observed data in the real world are used to estimate the hidden factors behind the phenomena that occurred in the system. The traffic demand estimation (OD matrix estimation) using iterative traffic simulations was described in Sect. 12.3. Furthermore, the value of obtaining data in order to reduce estimation errors and their uncertainty levels was discussed using the TCLP as an example.

Needless to say, the reliability of these approaches strongly depends on the reproducibility of the simulator. Guaranteeing simulator reproducibility for huge and complex social systems such as traffic systems is a hard task in itself, and efforts to improve simulation models should continue. However, even though models are imperfect, the authors believe that the process of understanding and designing a social system by utilizing both simulators and data is worth pursuing because there are few other alternatives.

Acknowledgements This work was supported in part by The Japan Society for the Promotion of Science (JSPS) KAKENHI Grant Numbers JP15H01785 and JP19H02377.

References

1. Abe, K., Fujii, H., Yoshimura, S.: Inverse analysis of origin-destination matrix for microscopic traffic simulator. Comput. Model. Eng. Sci. **113**(1), 71–87 (2017). https://doi.org/10.3970/cmes.2017.113.068
2. Abe, K., Yanai, M., Yamada, T., et al.: Optimization of traffic counting location considering demand uncertainty—evaluation by multi-agent-based traffic simulation. Trans. Jpn. Soc. Artif. Intell. **33**(6), D–I59_1–10 (2018). https://doi.org/10.1527/tjsai.D-I59. (in Japanese with English abstract)
3. Barceló, J. (ed.).: Fundamentals of Traffic Simulation. International Series in Operations Research & Management Science. Springer (2010). https://doi.org/10.1007/978-1-4419-6142-6
4. Bera, S., Rao, K.: Estimation of origin-destination matrix from traffic counts: the state of the art. Eur. Transp./Trasp. Eur. **49**, 3–23 (2011)
5. Chen, A., Pravinvongvuth, S., Chootinan, P., et al.: Strategies for selecting additional traffic counts for improving O-D trip table estimation. Transportmetrica **3**(3), 191–211 (2007). https://doi.org/10.1080/18128600708685673
6. Chootinan, P., Chen, A., Yang, H.: A bi-objective traffic counting location problem for origin-destination trip table estimation. Transportmetrica **1**(1), 65–80 (2005). https://doi.org/10.1080/18128600508685639
7. Du, Y., Jian, S.: Study on bi-objective traffic counting location optimization using ant colony method. In: 2009 International Workshop on Intelligent Systems and Applications, pp. 1–4 (2009). https://doi.org/10.1109/IWISA.2009.5073131
8. Ehlert, A., Bell, M.G., Grosso, S.: The optimisation of traffic count locations in road networks. Transp. Res. Part B: Methodol. **40**(6), 460–479 (2006). https://doi.org/10.1016/j.trb.2005.06.001
9. Fei, X., Mahmassani, H.S.: Structural analysis of near-optimal sensor locations for a stochastic large-scale network. Transp. Res. Part C: Emerg. Technol. **19**(3), 440–453 (2011). https://doi.org/10.1016/j.trc.2010.07.001
10. Fei, X., Mahmassani, H.S., Eisenman, S.M.: Sensor coverage and location for real-time traffic prediction in large-scale networks. Transp. Res. Rec. **2039**(1), 1–15 (2007). https://doi.org/10.3141/2039-01
11. Fujii, H., Yoshimura, S.: Precise evaluation of vehicles emission in urban traffic using multi-agent-based traffic simulator MATES. Comput. Model. Eng. Sci. **88**(1), 49–64 (2012). https://doi.org/10.3970/cmes.2012.088.049
12. Fujii, H., Sakurai, T., Yoshimura, S.: Virtual social experiment of tram railway extension using multi-agent-based traffic simulator. J. Adv. Comput. Intell. Intell. Inform. **15**(2), 226–232 (2011). https://doi.org/10.20965/jaciii.2011.p0226
13. Fujii, H., Uchida, H., Yoshimura, S.: Agent-based simulation framework for mixed traffic of cars, pedestrians and trams. Transp. Res. Part C: Emerg. Technol. **85**, 234–248 (2017). https://doi.org/10.1016/j.trc.2017.09.018

14. Hart, P.E., Nilsson, N.J., Raphael, B.: A formal basis for the heuristic determination of minimum cost paths. IEEE Trans. Syst. Sci. Cybern. **4**(2), 100–107 (1968). https://doi.org/10.1109/TSSC.1968.300136
15. Helbing, D., Tilch, B.: Generalized force model of traffic dynamics. Phys. Rev. E **58**(1), 133–138 (1998). https://doi.org/10.1103/physreve.58.133
16. Kitamura, R., Kuwahara, M. (eds.).: Simulation Approaches in Transportation Analysis. Operations Research/Computer Science Interfaces Series. Springer (2005). https://doi.org/10.1007/b104513
17. Lam, W., Lo, H.: Accuracy of OD estimates from traffic counts. Traffic Eng. Control **31**(6), 358–367 (1990)
18. Larsson, T., Patriksson, M.: Simplicial decomposition with disaggregated representation for the traffic assignment problem. Transp. Sci. **26**(1), 4–17 (1992). https://doi.org/10.1287/trsc.26.1.4
19. Levenberg, K.: A method for the solution of certain non-linear problems in least squares. Q. Appl. Math. **2**, 164–168 (1944)
20. Lundgren, J.T., Peterson, A.: A heuristic for the bilevel origin-destination-matrix estimation problem. Transp. Res. Part B: Methodol. **42**, 339–354 (2008). https://doi.org/10.1016/j.trb.2007.09.005
21. Marquardt, D.W.: An algorithm for least-squares estimation of nonlinear parameters. J. Soc. Ind. Appl. Math. **11**(2), 431–441 (1963). https://doi.org/10.1137/0111030
22. Matsuhashi, K., Kudoh, Y., Kamioka, N., et al.: A study on estimation method for transport CO_2 emissions by municipalities. Environ. Syst. Res. **32**, 235–242 (2004). https://doi.org/10.2208/proer.32.235. (in Japanese with English abstract)
23. McNally, M.G.: The four step model. Cent. Act. Syst. Anal. (2008). https://doi.org/10.1108/9780857245670-003
24. Wang, N., Gentili, M., Mirchandani, P.: Model to locate sensors for estimation of static origin-destination volumes given prior flow information. Transp. Res. Rec. **2283**(1), 67–73 (2012)
25. Yang, H., Iida, Y., Sasaki, T.: An analysis of the reliability of an origin-destination trip matrix estimated from traffic counts. Transp. Res. Part B: Methodol. **25**(5), 351–363 (1991). https://doi.org/10.1016/0191-2615(91)90028-H
26. Yang, H., Wang, Y., Wang, D.: Dynamic origin-destination estimation without historical origin-destination matrices for microscopic simulation platform in urban network. In: 2018 21st International Conference on Intelligent Transportation Systems (ITSC 2018), pp. 2994–2999 (2018). https://doi.org/10.1109/ITSC.2018.8569241
27. Yim, P.K.N., Lam, W.H.K.: Evaluation of count location selection methods for estimation of O-D matrices. J. Transp. Eng. **124**(4), 376–383 (1998). https://doi.org/10.1061/(ASCE)0733-947X(1998)124:4(376)
28. Yoshimura, S.: MATES: multi-agent based traffic and environment simulator—theory, implementation and practical application. Comput. Model. Eng. Sci. **11**(1), 17–26 (2006). https://doi.org/10.3970/cmes.2006.011.017

Chapter 13
Externalization of Unexplored Data with Data Origination: Case Analysis of Person-to-Object Contact Data During COVID-19 Pandemic

Teruaki Hayashi and Yukio Ohsawa

Abstract Various industries worldwide have been severely affected by the COVID-19 pandemic, highlighting the gaps between social systems and forcing major transformations of our lives. To understand and mitigate the phenomena related to the unprecedented danger of COVID-19, we have become acutely aware of the importance of data distribution, exchange, and sharing across fields; indeed, various data are published and used in decision-making processes. However, although many international organizations and companies have been publishing data and adopting relevant measures, data sharing regarding the question of what data are required for any purpose is insufficient; that is, data are principally provided by organizations who publish the data unilaterally; currently, data-related needs are not shared or leveraged. To address this issue, we introduce the concept of "data origination." Data origination is the act of designing/acquiring/utilizing data that considers the subjective knowledge and diversity of perspectives of humans, and that aims to elucidate and support this process. We also discuss a case study of data needs and unexplored data externalization conducted during the COVID-19 pandemic, based on data origination.

Keywords Data origination · Unexplored data · Data design · Data exchange · Knowledge sharing

13.1 Introduction

Recently, it has become possible to obtain data from various domains owing to the development of artificial intelligence (AI) and the global big data movement [5, 16]. In government fields, open data are published in a format facilitating secondary use,

T. Hayashi (✉) · Y. Ohsawa
Department Systems Innovation, School of Engineering, The University of Tokyo, 113-8656
7-3-1, Hongo, Bunkyo-ku, Tokyo, Japan
e-mail: hayashi@sys.t.u-tokyo.ac.jp

Y. Ohsawa
e-mail: ohsawa@sys.t.u-tokyo.ac.jp

© Springer Nature Switzerland AG 2023
Y. Ohsawa (ed.), *Living Beyond Data*, Intelligent Systems Reference Library 230,
https://doi.org/10.1007/978-3-031-11593-6_13

connected to several services, and generating value. In academic fields, language corpora and training data for image analysis are widely published and made available for secondary use. For instance, the National Institute of Informatics in Japan provides large-scale company-owned data in the form of the Informatics Research Data Repository for scientific use. In this climate, there are increasing expectations of data coordination across different domains, in combination with the societal need for digital transformation, e.g., emerging markets for data exchange creating value by combining heterogeneous data [15, 19].

Furthermore, various industries worldwide have been severely affected by the COVID-19 pandemic since the end of 2019, highlighting the gaps between systems and forcing major transformations of our lives. To understand and mitigate the phenomena related to the unprecedented danger of COVID-19, we have become acutely aware of the importance of data distribution, exchange, and sharing across fields. Data sharing and coordination across fields are essential to understanding and controlling unknown events. In fact, a project at Johns Hopkins University uses data from the American Centers for Disease Control and Prevention, the World Health Organization, and Chinese authorities to visualize the spread of COVID-19, and communicates this information to interested parties. In addition to local authorities such as the Tokyo Metropolitan Government and Cabinet Secretariat in Japan, private enterprises are publishing data and sharing a wealth of technologies and knowledge. Unquestionably, reusing and sharing data resources are important from the perspective of reducing costs. Although there are datasets provided online for free, such as open data, data acquisition is usually associated with high costs. Apart from the cost of data collection, the development, installation, and maintenance of equipment for data acquisition will incur costs. Considering this, making data available for reuse and shareable across organizations and fields is beneficial to society and human lives; however, there is one critical problem. If data are analyzed without considering the design concept of the creator of the data, wrong conclusions may end up being made. That is, if there is little background knowledge concerning the data (e.g., with what purpose did the creator build the data or what type of data they are), then any knowledge obtained from the data may be extremely biased. For instance, overfitting due to labeling errors has been observed in ImageNet, a large-scale database for images with training labels that is considered extremely useful in the field of image recognition [2]. Furthermore, Amazon recently stopped using their recruiting tool owing to the AI it employed, which discriminated against women [3]. In these examples, the use of AI was perceived to be the problem, but it became clear that there were issues with the quality of the data used for training. In present times, where things are predicated on big data, data designers, collectors, and analysts are often different entities. During the COVID-19 pandemic, large-scale public data have attracted attention; however, the purpose and the background behind acquiring them often remain unclear, and any understanding of the facts, and pertinent decision-making based on it, is far from adequate. American statistician Nate Silver said that "the number of reported COVID-19 cases is not a very useful indicator of anything— unless you also know something about how tests are being conducted,"[1] sounding a warning about viewing statistical data without a detailed understanding of the testing methods and data acquisition methods. As of this writing (August 2021), the COVID-19 pandemic still continues, and although many international organizations and corporations have been publishing data and implementing mitigation techniques, sufficient datasets that we want are seldom published and shared [13]. Equivalently, there is only unilateral data provision, and there is hardly any discussion about data creation for events that have not yet been observed or methodologies for supporting them.

This study proposes the concept of data origination as a solution to the problem that has manifested during the COVID-19 pandemic, namely, how to observe unknown events and convert them into data. The first half of the article describes the motivation of data origination with two techniques and outlines the insights into unexplored data, including potential data that have not yet been converted into data. The second half introduces the case analysis of externalization of unexplored data, using a survey dataset of human contact behavior with objects, which we gained from data origination workshops during the COVID-19 pandemic. The survey results reveal that, depending on the location, the types and numbers of objects that were touched differed, and the respective mean values of contact objects differed significantly. Although it is impossible to disinfect all objects and spaces, our findings will provide insights into human behavior and contact with objects. These findings are expected to contribute to the prioritization of disinfection during periods of widespread infection.

13.2 Data Origination

In fields such as the natural and social sciences, human interpretation is applied to data obtained through the observation of phenomena, or through experiments that try to understand the mechanisms underpinning those data. Examples for this include attempts to understand orbital periods by observing the movement of stars or earthquake mechanisms based on the occurrence of earthquakes and plate observations. Typically, two types of information are lost in the processes of observation and conversion into data and of knowledge acquisition. One is attribute information, which is not acquired in the process of observing and converting into data of certain events. This includes, for instance, information missed during the acquisition process owing to limitations of human recognition or sensor resolution. Another type is information not considered during the analysis of digitized data. As an example, in addition to temperature and amount of precipitation, weather data for a certain area will also include humidity and wind direction. However, when studying the relationship between store turnover and weather in that area, the link with temperature and amount of precipitation becomes apparent, and humidity and wind direction may be used for reference only and not included in the analysis. In this manner, many

---

[1] https://fivethirtyeight.com/features/coronavirus-case-counts-are-meaningless/.

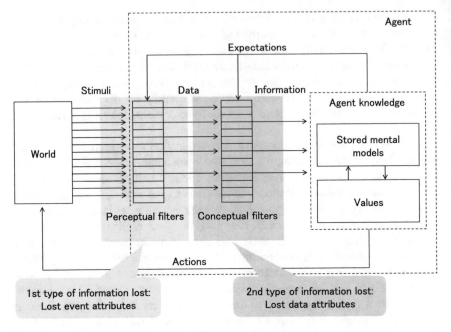

Fig. 13.1 Agent-in-the-world model (modified in part by the authors)

items are not used for analysis despite having been acquired as data. We will use the agent-in-the-world model [4] to explain these information losses and selections (Fig. 13.1). It models a process where stimuli that are emitted into the world by agents who observe (e.g., humans) are turned into knowledge. The world comprises phenomena from the natural world, including the society we live in, and various events emit numerous signals, i.e., information. To distinguish between the different types of information, we term information emitted by events as "stimuli," information obtained from data as "information," and the two types of lost information as "lost event attribute" and "lost data attribute," respectively.

In the agent-in-the-world model, agents observe stimuli emitted by the real world through a perceptual filter and convert them into data. For humans, this corresponds to our sensory organs; for machines, these are sensors. The data are then converted into information by passing them through a conceptual filter. The conceptual filter corresponds to a mechanism for interpreting data; it provides meaning to electrical signals or encoded data and converts it into knowledge. This is a mechanism where, in domains such as the natural and social sciences, human interpretation is applied to data obtained through the observation of events or through experiments that attempt to understand the mechanisms under-pinning those data.[2] Lost event attributes occur

---

[2] In this model, a mechanism is studied where knowledge further reinforces perceptual and conceptual filters, and influences the world as actions; however, a detailed discussion is omitted in this article.

during the process where events are converted into data through observation or experimentation. In the process of Fig. 13.1, losses occur during the process of stimuli being converted to data through a perceptual filter. Because data are an abstraction of events occurring in the real world and are reconstructed using character strings and code, substantial amounts of event attributes are lost in this first stage. For instance, as ultraviolet light cannot be detected with the human eye, these stimuli cannot be received as signals. In the context of sensors, not all stimuli can be converted into data owing to resolution limitations, which is the capacity of devices to measure or identify objects. However, this loss of event attributes is not necessarily a negative thing; for instance, when ascertaining road congestion, the model is simplified by abstracting moving cars into dots, facilitating the estimation of the degree of congestion. Here, information such as the number of people in the cars or car models may not be obtained; nonetheless, when information is unimportant for corroborating a hypothesis, it is usually not converted into data-in other words, event attributes are lost. Furthermore, in many cases, excess data will contain noise, making it difficult to extract the essence of events; thus, a moderate loss of event attributes makes subsequent analysis easier and increases the possibility of obtaining valuable data.

Loss of data attributes occur during the processes of analyzing, estimating, or predicting data and of corroborating a hypothesis. In Fig. 13.1, this corresponds to the process where data are converted into information through a conceptual filter. The need for this process lies in the complexity of events that have been converted into data and in the need for extracting, summarizing, and organizing the information further. If the essence of events under observation are appropriately converted into data, hypotheses can be corroborated using data analysis techniques. Once complex events are converted into data, statistical techniques can be applied to them even if some data are lost, or if some noise is contained within. Currently, many valuable, powerful tools have been proposed for phenomena characterized by variation that cannot be controlled by humans, or errors that occur through artificial measurement errors.[3] Currently, various data are published and made available online, including open data, and secondary use of data are widespread; however, we believe that major pitfalls to this practice are being overlooked. Data that are currently being shared have undergone the process of loss in the first stage; that is, they have been recorded after passing through the perceptual filter of data designers and collectors. Although data tend to considered as objective records of facts, it must be understood that the design intentions of data designers and collectors are clearly reflected in the data, and that data are strongly biased. Techniques, such as statistical modeling, are applied to the process where knowledge is obtained through acquired data, and they tackle lost data attributes, which are mainly information lost in the second stage; however, hardly any approach tackles lost event attributes at the first stage. To address this problem, the authors propose externalization of unexplored data through data origination [11]. Data origination is the act of data design/acquisition/utilization that considers the subjective knowledge and diversity of perspectives of humans and aims to elucidate

---

[3] Naturally, there is potential for human interpretation in the process where information is obtained from data, and the relevant analytical tools must be selected with care.

and support this process. Origination means beginning or origin, and is a human-centered approach that reverts to the act of human observation as the origin of data. Unexplored data signify a source of data that contains requests, imaginary parts, and events yet to be converted into data, that is, potential data of unobserved events. Unexplored data externalization through data origination indicates the process of people discovering events that they had not noticed existed and converting them into data. A concept similar to unexplored data is dark data [7], which refers to invisible, unrecorded, or hidden data. However, unexplored data covers both data for which people do not know what type of data can be obtained although they are aware of its existence as well as potential data for events of whose existence people are unaware. Therefore, it provides a granular definition that facilitates a more in-depth discussion than the case of dark data.

Data origination ensures the data quality, and aims to yield better solutions than those obtained thus far by utilizing useful data; however, it is not easy to establish a methodology to support and realize externalization of unexplored data. The reason for this is the sheer multitude and variation of information held within worlds or events that are being observed; humans are incapable of observing and recording everything. Even if highgrade sensing technologies make it possible to obtain data of a diversity exceeding the scope of human perception, the sensors are designed by humans. It is virtually impossible to obtain data that exceed the scope of human perception and design; moreover, because methods to observe events vary widely among research fields, the observation of unknown events is extremely difficult. One idea for externalizing unexplored data is to treat event attributes as a collection of variables and use the related collective intelligence for observations. Variables are the properties from which data are built; they are also termed elements or parameters. For instance, in the case of weather data, variables include "name of the region," "temperature," "humidity," and "weather," or in the case of physical data, variables would correspond to "height," "weight," "age," "blood type," etc. When using the minimum attribute unit of acquired data as variables, what type of variable sets are to be obtained as data is the key question to be considered when designing unexplored data. In the next section, we introduce two support tools for unexplored data externalization based on a collective intelligence approach that focuses on variables of data.

13.3 Support Tools for Externalizing Unexplored Data: TEEDA and Variable Quest

13.3.1 TEEDA

The provision of data portals and platform services has led to an increase in opportunities for users to learn about data held by data owners and providers. However, information about the types of data users want and their purpose is being shared

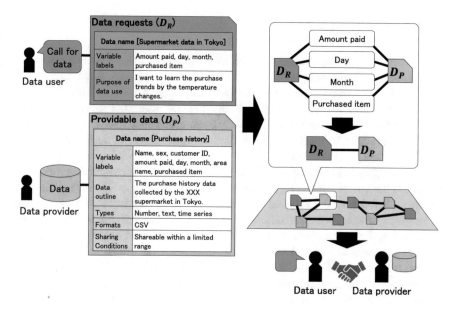

Fig. 13.2 Process of collecting and matching data requests and providable data in TEEDA

insufficiently. Thus, data owners are unable to obtain the required types of data, and the risk exists that the sites only provide data that do not meet users' requirements. Treasuring Every Encounter of Data Affairs (TEEDA) is a tool that was developed to promote communication between, and matching of, data owners and users by supplying requests for data wanted by users to the market, in a data marketplace for data exchange between different industries [12]. Figure 13.2 shows that data owners register information as "providable data" on TEEDA, and that data users submit their needs in terms of the data structure they require as "data requests" (the combination of both is termed a "data item"). The collected data items are processed using matching algorithms and visualized, thereby promoting data exchange between data owners and users.

TEEDA[4] is an online tool that was developed to support data coordination and collaboration across different fields, and it functions as a support tool for unexplored data externalization in the context of data origination. Data requests in TEEDA correspond to data needs, i.e., unexplored data, which are potential data for unobserved events. In general, data needs are potentially described in various formats, and not easily discussed on an equal footing. Three description entries have been set for data requests in TEEDA: name of the desired data, variable set, and purpose of use. These allow for descriptions of various data needs in natural language, which enables externalization of lost event attributes, i.e., the data to be obtained as a combination of the relevant variables and their purpose.

---

[4] https://teeda.data-marketplace.org/.

13.3.2   Variable Quest

Sometimes, after having designed and collected data, users may realize the benefits of including an additional, different type of data. For example, forgetting to add a question about respondents' nationality to a questionnaire when obtaining the attributes of attendees of an event will significantly affect the subsequent result analysis. Particularly for events that can only be observed once, such as an earthquake or the Olympics, it is important to consider the type of data to design the type of variables to use prior to the event.

Variable Quest[5] is a system with databases and algorithms for estimating variables, which are the constituting components of data [9, 10]. Variable Quest models data similarity (Model 1) and variable co-occurrence (Model 2); through the provision of variables of unknown data outlines in text, it generates variable sets that can be contained in those data. Data similarity is a quantification of how similar data are to other data, and is a model that assumes that, if the data type and structure are similar, then the obtained variables are also similar (Fig. 13.3a). However, variable co-occurrence is a quantification of features that occur simultaneously in data with certain variable pairs. For instance, "longitude" and "latitude" will occur frequently in data that represent locations; meanwhile, "weight," "height," and "age" are likely to occur in medical check-up data (Fig. 13.3b). Hence, if a question about annual population trends in Japan is entered into Variable Quest, variable sets that may be contained within the data including "number of births," "number of deaths," "number of immigrants," and "population," are generated. This makes it possible to estimate related variable sets that will be contained in the data, based on the text of data outlines whose variables are unknown.

The benefit of using Variable Quest for unexplored data externalization is that, by providing the outlines of events that have not yet been converted into data as text, variable sets related to these events can be obtained from the collective intelligence related to the data, even if no keywords related to the data or variables are contained. Thus, Variable Quest is a support tool for questions regarding the type of stimuli from which the data must be obtained as variable sets by verbalizing the event to be observed. Equivalently, similar to TEEDA, it plays a role in supplementing lost event attributes.

13.4   Unexplored Data Externalization and Analytical Process

This section introduces a case where TEEDA and Variable Quest were used to generate person-to-object contact data during the COVID-19 pandemic. The research team that we were part of developed a workshop technique to promote data collabo-

---

[5] https://variable-quest.data-marketplace.org/.

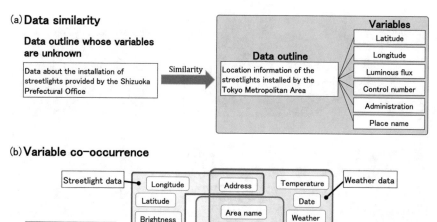

Fig. 13.3 a Model 1: Pairs of data items for which the similarity of the data outlines is high are similar. **b** Model 2: Pairs of data items that have the same set of variable labels are similar

ration through communication between participants from different fields who were involved in the utilization of data. However, owing to the global spread of COVID-19 since the end of 2019, it became increasingly difficult to host in-person workshops; for comparison, while the authors hosted 38 in-person workshops in 2019, they only did so once in 2020. Against this background, by using third-party video call systems such as Zoom or Microsoft Teams, we constructed an environment for data collaboration workshops that were conducted fully online. Hence, we conducted 13 TEEDA workshops, and obtained 660 data items from 163 participants in 2020.

After the workshops, upon detailed examination of the providable data and data requests entered among the research team, there was clearly a need to investigate places and objects with high frequency contact, and to consider guidelines for the location of sanitizer stations, objects to be sanitized with priority, and for contact behaviors on locations and transport (Fig. 13.4). There was little data about mitigation measures for infection depending on contact with people and objects (people touching/being touched by something) that resulted in effective infection mitigation measures, and they did not go beyond the recommendations of disinfecting hands through hand washing and sanitization, disinfecting objects, etc. Because people normally touch many objects, from personal items (such as keyboards and smartphones) to objects that are touched by high numbers of unspecified people (such as doorknobs, strangers on trains, and cash), it is realistically impossible to sanitize all objects in any given location. Thus, we believed that converting the relationship between behavior and touched objects into data and analyzing them helps ascertain the actual status of human behavior and contact with objects; moreover, it contributes to the proposals for prioritizing disinfection in the period of spreading infection.

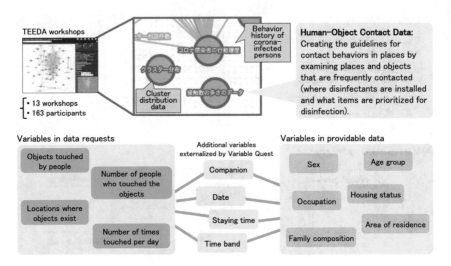

Fig. 13.4 Example of design process for unobserved data using TEEDA and variable quest

Our team discussed a hypothesis to obtain valuable analytical data using the following variable sets constituting these data requests as origin: "objects touched by people," "number of people who touched the objects," "locations where objects exist," and "number of times touched per day." Consequently, we concluded that an understanding of the actual situation concerning human behavior and contact with objects could be achieved by adding variable sets constituting providable data, i.e., "sex," "age group," "occupation," "housing status," "family composition," and "area of residence," in addition to sets that were externalized using Variable Quest, i.e., "companion," "date," "staying time," and "time band" (Fig. 13.4).

Fifteen locations and vehicles were selected, including locations such as "restaurants" and "gyms," for which infection clusters were reported in April 2020, and those that are part of people's daily lives such as "supermarket/convenience store (CVS)," "hospital," and "train." A survey was sent out to 1,288 subjects living in the Tokyo Metropolitan Area and Kanagawa Prefecture in Japan, asking respondents about their behaviors and touched objects while outside, between December 3-5, 2020.

The results of the survey showed that 7,317 objects (689 types) were touched in 15 types of locations/vehicles by 1,260 respondents returning valid answers. The objects most touched by respondents were doors, chairs, and shopping-related accessories, such as baskets, elevators, and cash at cash registers (Fig. 13.5a). Because elevated levels of SARS-CoV-2 have been detected on trashcan and door handles of commercial establishments, such as food stores [8], building and room doors were touched by many unspecified people, including those infected, on numerous occasions. Considering the samples taken from toilet doorknobs in hospitals that tested PCR positive [6, 17], we inferred that focus should be placed on disinfecting doors; however, despite the high contact frequency for chairs and shopping baskets, they

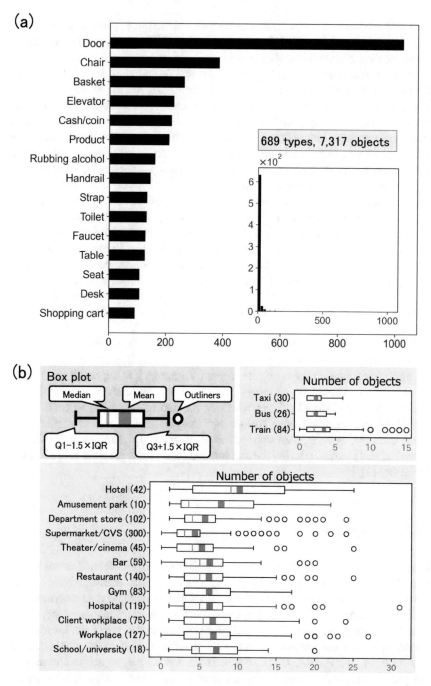

Fig. 13.5 a Distribution of 15 most touched objects (bins: 50). **b** Box plots showing the number of objects that were touched at each location and per vehicle type. The numbers in parentheses are the numbers of users

were scarcely referred to in previous research. Furthermore, because virus survival rates on cash are high [18], it is evident that attention must be paid to giving and receiving cash and associated behaviors.

We also analyzed this according to location and vehicle (Fig. 13.5b); for instance, the number of hotel users was low at 42 respondents, but the average number of touched objects per person was 10.7 objects over 132 object types, revealing high numbers. The results suggested the possibility that a higher number of diverse objects were touched during long-distance journeys or holidays compared with those in daily life. Furthermore, supermarkets/CVS had many users at 300 respondents, but the average number of objects touched was 4.45 objects per person for more than 218 object types, revealing that fewer objects were touched than in other locations. It was believed that this was because people at supermarkets/CVS had specified objectives of purchasing goods, and hardly any unnecessary objects were touched.

Moreover, we found that, for locations where food- and drink-related objects were served, such as restaurants, bars, and theaters/cinemas, having lunch in dining halls were reported for client workplaces, workplaces, schools/universities, hotels, and hospitals, whereas eat-in meals were reported for department stores. The removal of masks is expected at these locations; therefore, effective ventilation is required, and hands and objects with frequent contact occurrences should be thoroughly disinfected.

Previous studies have noted the remnants of viruses on doorknobs and toilets. However, the general dynamics of these contact numbers indicated that the percentage of people encountering these objects is small. Although it is impossible to disinfect all objects and spaces, our findings in this survey and analysis provide insights into human behavior and contact with objects, and are expected to contribute to prioritizing disinfection during periods of widespread infection. For more detailed analysis results, we refer interested readers to reference [14]. Moreover, The datasets generated and analyzed during the current study are available in the Kaggle dataset repository [1].

13.5 Conclusion

Based on our awareness of the current situation where data-related needs are not shared sufficiently, we explained our proposed concept of data origination in this article, and outlined a case study of data needs and unexplored data externalization conducted during the COVID-19 pandemic. For the case study, human-object contact behaviors during the COVID-19 pandemic were converted into data; however, we intend to proceed with data design and acquisition that meet other data needs obtained with TEEDA and Variable Quest. Furthermore, the main approach in current technologies for supporting data origination involves collective intelligence, which uses information relating to the data needs of various users and data obtained in the past. The clarification of unknown events that nobody is aware of or has observed, and of events that cannot be observed, is another motivation for data origination and unexplored data. The development of new techniques through post-collective intelligence is a problem that must be solved in future. It has long been said that data are the new oil of the 21st century; however, there is a surprising lack of understanding of events occurring in our world, and only a small part is understood and converted into data. There are several unknown events that have yet to be observed, which have just not been recorded or recognized. In the future, we hope that research into, and supporting technologies for, data origination will encourage attention to such unknown events. Furthermore, we hope that this will contribute toward the use and application of valuable data and the development of solutions.

Acknowledgements This study was supported by JSPS KAKENHI (JP19H05577 and JP20H02384), the "Startup Research Program for Post-Corona Society" of Academic Strategy Office, School of Engineering, the University of Tokyo, the "COVID-19 AI and Simulation Project" of the Office for Novel Coronavirus Disease Control, Cabinet Secretariat, Government of Japan, and the MEXT Quantum Leap Flagship Program (MEXT Q-LEAP) under Grant JPMXS0118067246. We thank Editage for providing English language editing.

References

1. (2020) Person-to-object contact dataset: actual conditions of contact behaviors with objects during the covid-19 pandemic. https://www.kaggle.com/teruakihayashi/person-to-object-contact-dataset
2. Beyer, L., Hénaff, O.J., Kolesnikov, A., et al.: Are we done with imagenet? (2020). arXiv:2006.07159
3. Black, J.S., van Esch, P.: Ai-enabled recruiting: what is it and how should a manager use it? Bus. Horiz. **63**(2), 215–226 (2020)
4. Boisot, M., Canals, A.: Data, information and knowledge: have we got it right? J. Evol. Econ. **14**(1), 43–67 (2004)
5. Chui, M., Manyika, J., Miremadi, M., et al.: Notes from the AI Frontier: Applications and Value of Deep Learning. McKinsey Global Institute, London (2018)
6. Guo, Z.D., Wang, Z.Y., Zhang, S.F., et al.: Aerosol and surface distribution of severe acute respiratory syndrome coronavirus 2 in hospital wards, Wuhan, China. Emerg. Infect. Dis. **26**(7), 1586 (2020)
7. Hand, D.J.: Dark Data: Why What You Don't Know Matters. Princeton University Press, Princeton (2020)
8. Harvey, A.P., Fuhrmeister, E.R., Cantrell, M.E., et al.: Longitudinal monitoring of SARS-COV-2 RNA on high-touch surfaces in a community setting. Environ. Sci. Technol. Lett. **8**(2), 168–175 (2020)
9. Hayashi, T., Ohsawa, Y.: Variable quest: network visualization of variable labels unifying co-occurrence graphs. In: 2017 IEEE International Conference on Data Mining Workshops (ICDMW), pp. 577–583. IEEE (2017)
10. Hayashi, T., Ohsawa, Y.: Inferring variable labels using outlines of data in data jackets by considering similarity and co-occurrence. Int. J. Data Sci. Anal. **6**(4), 351–361 (2018)
11. Hayashi, T., Ohsawa, Y.: Data origination: human-centered approach for design, acquisition, and utilization of data. In: International Conference on Soft Computing and Pattern Recognition, pp. 85–93. Springer, Berlin (2020a)
12. Hayashi, T., Ohsawa, Y.: Teeda: an interactive platform for matching data providers and users in the data marketplace. Information **11**(4), 218 (2020)

13. Hayashi, T., Uehara, N., Hase, D., et al.: Data requests and scenarios for data design of unobserved events in corona-related confusion using teeda. In: 2020 IEEE International Conference on Big Data (Big Data), pp. 4456–4461. IEEE (2020)

14. Hayashi, T., Hase, D., Suenaga, H., et al.: The actual conditions of person-to-object contact and a proposal for prevention measures during the covid-19 pandemic (2021). medRxiv

15. Liang, F., Yu, W., An, D., et al.: A survey on big data market: pricing, trading and protection. IEEE Access **6**, 15,132–15,154 (2018)

16. Manyika, J., Chui, M., Brown, B., et al.: Big data: the next frontier for innovation, competition, and productivity. McKinsey Global Institute (2011)

17. Ong, S.W.X., Tan, Y.K., Chia, P.Y., et al.: Air, surface environmental, and personal protective equipment contamination by severe acute respiratory syndrome coronavirus 2 (sars-cov-2) from a symptomatic patient. JAMA **323**(16), 1610–1612 (2020)

18. Riddell, S., Goldie, S., Hill, A., et al.: The effect of temperature on persistence of SARS-CoV-2 on common surfaces. Virol. J. **17**(1), 1–7 (2020)

19. Stahl, F., Schomm, F., Vossen, G.: Data marketplaces: an emerging species. In: Frontiers in Artificial Intelligence and Applications, pp. 145–158. IOS Press (2014)

Index

© Springer Nature Switzerland AG 2023
Y. Ohsawa (ed.), *Living Beyond Data*, Intelligent Systems Reference Library 230,
https://doi.org/10.1007/978-3-031-11593-6

Printed in the United States
by Baker & Taylor Publisher Services